D1192133

W.R. LAIRD

The Unfinished Mechanics of Giuseppe Moletti

An Edition and English Translation of His *Dialogue on Mechanics*, 1576

UNIVERSITY OF TORONTO PRESS
Toronto Buffalo London

Toronto Buffalo London
Printed in Canada

ISBN 0-8020-4699-1

Printed on acid-free paper

Canadian Cataloguing in Publication Data

Moletti, Giuseppe, 1531–1588
The unfinished mechanics of Giuseppe Moletti : an edition and English translation of his Dialogue on Mechanics (1576)

Includes bibliographical references and index.
ISBN 0-8020-4699-1

1. Mechanics – Early works to 1800. I. Laird, Walter Roy, 1950– .
II. Title.

QA804.M64 2000 531 C99-931367-3

Some material reused and adapted from 'Giuseppe Moletti's "Dialogue on Mechanics" (1576),' *Renaissance Quarterly* 40 (1987): 209–23.

Some material reused and adapted from 'The Scope of Renaissance Mechanics,' *Osiris*, 2nd series 2 (1986): 43–68, by permission of the University of Chicago Press.

University of Toronto Press acknowledges the financial assistance to its publishing program of the Canada Council for the Arts and the Ontario Arts Council.

This book has been published with the help of a grant from the Humanities and Social Sciences Federation of Canada, using funds provided by the Social Sciences and Humanities Research Council of Canada.

University of Toronto Press acknowledges the financial support for its publishing activities of the Government of Canada through the Book Publishing Industry Development Program (BPIDP).

To the memory of

Stillman Drake

un secondo Galileo

Contents

Plates

Figures

Preface

This book has been so long in the making that at times I feared it might suffer the same fate as Moletti's mechanics. That it appears at last is due in no small measure to the help and encouragement that I received from a number of people and institutions. For their financial support I should like to thank the Social Sciences and Humanities Research Council of Canada, the Mellon Foundation, the University of Toronto, and Carleton University, from all of whom I received at various times scholarships and grants to aid my travel and research. I should also like to thank the Biblioteca Ambrosiana in Milan and the Ambrosiana Microfilm Collection at the Medieval Institute, University of Notre Dame, especially Marsha Kopacz and Louis Jordan, who provided me with microfilms and photographs of Moletti's papers. My thanks go also to the Vatican Library, to the Archivio di Stato in Mantua, and to the Biblioteca Civica in Belluno for microfilms and photocopies, and to the Thomas Fisher Rare Book Library at the University of Toronto for photocopies and microfilms from their Aristotle collection and from Stillman Drake's incomparable collection of Galileiana. Some of the material in the introduction has appeared earlier in articles published in *Osiris* and in *Renaissance Quarterly*; I should like to thank their editors for permission to use it here. Bert Hansen, Bert Hall, Albert Van Helden, Thomas B. Settle, and Nicholas Adams all gave me help and suggestions at various stages of this work, only some of which can be acknowledged in notes. I am especially grateful to Francesco Guardiani, Department of Italian Studies, University of Toronto, who read through the Italian text and the English translation and saved me from a number of embarrassments. Any shortcomings that remain are, of course, entirely mine. Finally, the dedication inadequately repays an old debt to the incomparable teacher and scholar who first got me started on sixteenth-century mechanics.

THE UNFINISHED MECHANICS OF GIUSEPPE MOLETTI

Introduction

Cose tutte imperfette et senza risolutione certa
– Milan, Biblioteca Ambrosiana MS. S 100 sup., f. 295r

It is a melancholy fact that most of works begun by Giuseppe Moletti were left unfinished at his death. His extant papers brim with the hopeful beginnings of numerous treatises and commentaries on all the diverse subjects that were proper to the *mathematicus* of the sixteenth century, yet few of these are complete in manuscript, and fewer still were printed. Only four separate works were ever published under his name: an introduction to geography, which was printed with a translation of Ptolemy's *Geography*; an *Ephemerides*; a book of astronomical tables; and a treatise on the reform of the calendar.[1] A fifth work, a geographical-political treatise on the supremacy of the Spanish monarchy, was printed under another's name shortly after Moletti's death.[2] These few printed works hardly do justice to the wide mathematical interests revealed in Moletti's unpublished papers. There one finds the expected introductions to and annotations on the school texts he lectured on at the University of Padua at the end of his career: Euclid's *Elements*, the optics of Alhazen and Witelo, Sacrobosco's *Sphere*, and pseudo-Aristotle's *Mechanical Problems*. Among his papers are proposals for commentaries on the *Sphere* of Theodosius, on Archimedes' *Sphere and Cylinder*, and (most interestingly) on Copernicus's *De revolutionibus*. He also left several philosophical works on the nature of mathematics and on mathematical certainty, one of which is actually complete, and numerous proposed works, including an *Ephemerides* based on Copernicus, an introduction to astrology, a work on the mobile sphere, and one on arithmetic. In addition, his papers contain fragments on more popular sub-

jects: games with the compass, a description of the *radio geometrico* (an instrument for measuring distance), and a work each on the horologium, on practical measurement, on fortification, and on practical perspective. His literary remains also include numerous personal letters, many treating mathematical topics, and a dizzying array of miscellaneous notes on all manner of mathematical things.[3] Taken together, all these materials suggest a restless mind that ranged from the highest abstractions to the humblest practical applications, a life devoted mainly to teaching others, and great plans thwarted by poor health and an early death.

The edition and translation that follows is of one of those works begun so hopefully but abandoned midway. On 1 October 1576, 'with the altitude of the sun 17 degrees, the hour 14:33' – ever the mathematician, he customarily dated his sheets thus – Moletti began an Italian *Dialogue on Mechanics,* mechanics being one of the sciences proper to the mathematician and one on which he would later lecture at the University of Padua. Like those later lectures, the *Dialogue* was based on an ancient text then recently recovered and translated into Latin, the pseudo-Aristotelian *Mechanical Problems,* perhaps the most influential work on mechanics in the sixteenth century. But the *Dialogue* was intended to do much more than merely explain the Aristotelian text. Rather, Moletti's twofold purpose was, first, to establish the science of mechanics on rigorous, mathematical principles, and second, to extend mechanics into topics until then treated mainly in natural philosophy. In both these purposes Moletti was working within the tradition of mechanics that had recently developed around the *Mechanical Problems* and that was transforming the very status of the science.

Largely under the influence of the *Mechanical Problems,* mechanics was increasingly seen in the course of the sixteenth century as a theoretical, mathematical science rather than as a purely practical art, and identical in status to astronomy, optics, and harmonics. Such sciences were known since the Middle Ages as middle or subalternated sciences, since they dealt with natural things but took their principles from the higher sciences of geometry and arithmetic. In the First Day of the *Dialogue on Mechanics* Moletti attempted in a series of rigorous mathematical theorems to establish the Euclidean foundations of the principle of Aristotelian mechanics, and thus to realize in fact the status of mechanics as a mathematical science subalternated to geometry.

In the unfinished Second Day, Moletti attempted to extend mechanics into realms long the province of natural philosophy. Taking his start from Aristotle's *De motu animalium,* he tried to sort out the relations

between the principles of motion – motive powers, resistances, and the immobile – and to compare the speeds of falling bodies and discover the cause of their acceleration. Here he was less successful, though sometimes very insightful: he recognized, for instance, that heavy bodies resist motion to the same extent that, once moving, they continue it, and that the force of impact of a moving body depends not only on its size and speed but also on the resistance encountered. He also argued that falling bodies of the same shape fall at the same speed no matter what their materials or weights, and he claimed to have tested the result many times – all this while Galileo was still a boy. The *Dialogue* breaks off during an inconclusive discussion of the cause of the acceleration of falling bodies, a question that lay well beyond the scope of Moletti's mechanics.

Moletti was concerned with mechanics not only as a theoretical, mathematical science, but also for the practical importance of its applications. When he wrote the *Dialogue*, he was employed by Duke Guglielmo Gonzaga of Mantua as mathematics tutor to the young prince Vincenzo, and the work takes its form and part of its purpose from these circumstances. First, it was written in Italian, not Latin, and so was meant for a courtly rather than an academic audience; second, it is a fictional dialogue meant to persuade, not a didactic lecture or commentary; and finally, it appealed frequently to practical observations and experiences of everyday things. The interlocutors, identified only by initials, were not professors or academics: the mechanical adept in the *Dialogue* is the Duke of Mantua himself, while his companion is a visiting noble and military officer. Implicit in this arrangement is a plea for the usefulness of mechanics to princes and soldiers, and for the value of those who are expert therein.

Unlike Galileo after him, Moletti did not conceive of a separate science of motion to be founded on mathematical principles, as were astronomy, optics, and – of course – mechanics, and he probably lacked both the mathematical sophistication and the experimental temperament to create one. Nevertheless, the unfinished *Dialogue on Mechanics* offers an intriguing insight into the *practice* of mechanics (if I may speak this way of a theoretical science) at a princely court of the late Renaissance – the attempt to establish its theoretical underpinnings, to extend it into natural philosophy, and to present its practical applications. In the Introduction to the edition and translation that follows, I shall discuss first the meaning and scope of mechanics in the Middle Ages and Renaissance and the main traditions by which it was conveyed. The

most important of these in the sixteenth century, and certainly for Moletti, was the tradition of the pseudo-Aristotelian *Mechanical Problems*, which was translated by and commented on by a number of his sixteenth-century predecessors and contemporaries. Against this background I shall then sketch Moletti's life and works, with special attention to his earlier *Discourse on How to Study Mathematics* (*Discorso che cosa sia matematica*), from which he borrows extensively in the *Dialogue*, and his later *Lectures on the Mechanical Problems*. There then follows a summary of the *Dialogue* itself, in which I shall try to lay bare its logical and mathematical structure and suggest its main sources and influences. Finally, this Introduction will conclude with the career of mechanics and motion after Moletti, culminating in Galileo, where I shall venture a few comparisons between the two.

I Mechanics in the Middle Ages and the Sixteenth Century

Before the sixteenth century, the topics that now make up classical or Newtonian mechanics were distributed among a variety of sciences and arts in a way that corresponds only roughly to the modern divisions of statics, dynamics, and kinematics.[4] The general principles of motion and change were the province of *physica* or *philosophia naturalis* (natural philosophy), the particular branches of which treated the various special kinds of motion and change. What historians now often refer to as medieval dynamics and kinematics comprised topics usually treated in natural philosophy, although these topics could also arise in contexts as different as logical or theological works. In addition, from the mid-thirteenth century on, independent treatises on motion (*de motu*) began to appear, which isolated various problems of dynamics and kinematics for special treatment. In the fourteenth century all these genres applied to problems of motion the medieval theory of ratios and proportions and the intension and remission of forms.[5]

Topics associated with the balance and lever made up the medieval science of weights (*scientia de ponderibus*), which was considered a branch of mathematics rather than a part of natural philosophy. This science enjoyed a tradition separate from that of natural philosophy, centring on the works of the thirteenth-century Jordanus de Nemore, although Archimedes' statical works and their Arabic derivatives were also extant, if not as widely known. The science of weights was not statics in the strict, Archimedean sense, for it was based on principles involving motion and occasionally treated problems concerned with

moving bodies.[6] Nor did it treat any simple machine beyond the lever and (sometimes) the inclined plane; in fact, there does not seem to have been in the Middle Ages any theoretical treatment of the wheel and axle, pulley, wedge, or screw.

The construction and use of machines, however, belonged to the various mechanical arts (*artes mechanicae* or *sellulariae* or *machinativae*), which included architecture, metalworking, weaving, and the like. The transmission of such practical knowledge was without doubt largely oral, although some books on machines, especially military machines, are extant.[7] Even the works of the later artist-engineers, the outstanding examples being those of Leonardo da Vinci, are long on the design and application of machines but short on their theory.

Medieval thinkers, unable to identify any known work of Aristotle's with the science he referred to as mechanics, were forced to conjecture about the nature of that science. They did not, however, consider mechanics a speculative science, as they did optics, astronomy, and harmonics, but usually identified it with the set of mechanical or sellularian arts that employed mathematics. Mechanics was occasionally identified with engineering (*scientia de ingeniis*),[8] but never with the science of weights, with Archimedean statics, or with Aristotelian dynamics. One of the effects of the reintroduction of the pseudo-Aristotelian *Mechanical Problems* in the sixteenth century was to elevate mechanics to the status of a theoretical, intermediate science and to apply to it the theory of subalternation elaborated in the Middle Ages and still actively discussed in the sixteenth century. As I shall argue below, Moletti's purpose in the First Day of the *Dialogue* was to establish mechanics on Euclidean foundations and thus to realize in fact its subalternation to geometry.

All of the traditions I mentioned above – the natural philosophical, the Jordanian and Archimedean, and the practical – continued into the sixteenth century. But this century also saw the introduction and assimilation of additional ancient mechanical and technological works, due largely to the efforts of the humanists of the preceding century in recovering and copying the texts. Among the works so recovered was the pseudo-Aristotelian *Mechanical Problems*. Although now thought to have been written in the fourth century BC by Strato, a student of Aristotle's,[9] in the sixteenth century the *Mechanica* or *Problemata mechanica* was widely, if usually with reservations, attributed to Aristotle. The text was apparently unknown throughout the Middle Ages, although it seems to have had considerable, if indirect, influence on Jordanus.[10] In the fifteenth and sixteenth centuries, however, the work enjoyed a resurgence

of interest, attracting the attention of numerous and diverse commentators, including humanists, philosophers, mathematicians, and engineers.[11] Under its aegis they developed a program for mechanics that incorporated elements not only from the Jordanian and Archimedean traditions and the practical tradition of machines, but also from peripatetic natural philosophy. At the same time, these commentators took great care to distinguish this new science of mechanics from practical mechanics on the one hand and from natural philosophy on the other, placing it among such mathematical sciences as astronomy and optics.

In general, mechanics in the sixteenth century was distinguished from other sciences by four main characteristics: first, it was a theoretical science rather than a manual art; second, it was mathematical, although its subject was natural; third, it concerned motions and effects outside of or even against nature; and fourth, it produced them for human ends. These four characteristics, which together define the scope of renaissance mechanics, derive from the opening paragraph of the *Mechanical Problems*. The purpose of the work, according to that paragraph, is to explain marvels that occur against nature and that are produced by skill for the benefit of man, in contrast to marvels that occur in accordance with nature. Implicit here is the contrast between mechanics and natural philosophy, which explains only marvels and other events occurring in accordance with nature and which does not consider their usefulness to man. For nature, as the text continues, acts only in one way, but what benefits man acts in many ways. Such marvels against nature occur when the lesser masters the greater, when a lighter weight lifts a heavier, and all similar problems termed mechanical.[12]

Mechanical problems do, however, involve natural philosophy to some extent, for, as the text of the *Mechanical Problems* continues, they share in both mathematical and natural speculations; that is, 'the how' (*τὸ ὥς*) of mechanical problems is revealed through mathematical speculations while 'the about-what' (*τὸ περὶ ὅ*) is revealed through natural speculations.[13] This notion derives from a passage in the *Physics*, where Aristotle characterizes optics, astronomy, and harmonics as 'the more physical (or natural) of the mathematical sciences.' To these three sciences he added a fourth, mechanics, both in a passage in the *Metaphysics* and in his extensive discussion of such sciences in the *Posterior Analytics*.[14] The chief characteristic of these four sciences was that they were 'under' a purely mathematical science (either arithmetic or geometry), from which they derived their causes and demonstrations, while their subject matter was in some way natural or sensible. Medieval peripatet-

ics sometimes called optics (*perspectiva*), astronomy, and harmonics (*musica* or *harmonica*) the 'middle sciences' (*scientiae mediae*), since they viewed them as occupying an intermediate position between mathematics and natural philosophy. For them, the middle sciences were 'subalternated' (*subalternatae*) to mathematics, while their subjects were 'determined,' 'contracted,' or applied to some kind of sensible or natural matter.[15]

Immediately after this introductory paragraph of the *Mechanical Problems* there follows a section that claims that mechanical marvels are to be explained by the marvellous properties of the circle, which are then enumerated. Chief among these properties is that a point on a rotating radius moves at a speed that is in accordance with its distance from the centre. This is what I call the principle of circular movement, and this is the mechanical principle that Moletti in the First Day of the *Dialogue* attempts to ground in Euclidean geometry. On this principle depend the properties of the balance (Latin *libra*), and on the balance, in turn, depend the properties of the lever (Latin *vectis*). Finally, almost all other mechanical movements depend on the lever.[16] This chain of dependence places the balance ahead of the lever epistemologically and serves to distinguish them, for the balance was considered an analytic instrument rather than a machine, while the lever was considered a machine that applies the principles of the balance to produce useful work. They were not distinguished, it should be emphasized, by the balance's being static and the lever dynamic, for they both depend on the same fundamental principle of circular motion. The rest of the text of the *Mechanical Problems* consists of some thirty-five problems to which these principles are applied, problems concerning the balance, lever, pulley, wedge, oars, rudders, and a variety of miscellaneous topics, including the breaking of beams and projectile motion.[17]

As I mentioned above, in the sixteenth and early seventeenth centuries a variety of commentators brought their varied interests and expertise to bear on the *Mechanical Problems*. These I have categorized and discussed in some detail elsewhere; here I should like to review only those that influenced Moletti's conception of mechanics and that might shed some light on his treatment of it in the *Dialogue*.[18]

The *Mechanical Problems* was first made available to Latin readers in the early sixteenth century by the efforts of several humanists, who, in the course of their study of Greek and Greek philosophical works, translated, paraphrased, and commented on the text. Although their interest in the *Mechanical Problems* was more philological and philosophical than

mathematical, they helped to establish mechanics as a legitimate mathematical science within the Aristotelian scheme of knowledge, and they introduced the work to a wide readership.

The first Latin translation of the *Mechanical Problems* was made by the Venetian humanist Vittore Fausto (1480–1551?) and published in 1517.[19] Only occasionally cited by later writers, including Moletti, Fausto's translation was largely supplanted by one made only a few years later by another Venetian, Niccolò Leonico Tomeo (1456–1531), who published his translation of and annotations on the *Mechanical Problems* in Venice in 1525.[20] Leonico was professor of philosophy at the University of Padua from 1497 to about 1509, although he also taught Greek in Venice from 1504 to 1506. From 1521 on he taught privately in Padua. Leonico's enthusiasm for Aristotle in Greek is witnessed by his epitaph, which reads: 'He was the first at Padua to teach the books of Aristotle in Greek.' Moletti described him as 'very learned in Greek and an excellent philosopher of peripatetic philosophy here at the University of Padua.' Leonico's translation of the *Mechanical Problems* was reprinted numerous times after the first edition of 1525, both with the works of Aristotle and as the text accompanying later commentaries, and it became the version most often used in the sixteenth century. In addition to his *Annotationes in Mechanicas quaestiones*, Leonico translated and commented on Aristotle's *Parva naturalia* and book 1 of the *De partibus animalium*, and he paraphrased the *De motu animalium* and the *De ingressu animalium*. The last two paraphrases were printed with the Venice (1525) edition of the commentary on the *Mechanical Problems*, which is perhaps why, of all Aristotle's natural works, Moletti found it convenient to draw from these two works in the Second Day of his *Dialogue on Mechanics*.[21]

In his commentary on the *Mechanical Problems* Leonico first notes that Aristotle called what is done by art 'beyond nature' ('praeter naturam') because art must often transgress the laws of nature to effect things for the benefit of man. Mechanics, according to the Greeks, he continues, was that part of the art of building that used machines. While its subject matter is natural, since machines are made of natural materials, its mode and power of working ('modum operandique vim') are mathematical, for weights and measures are abstracted from the natural material in which they are found in order to reveal their 'principles of design' ('rationes formarum').[22] Leonico thus accepts the main features of Aristotelian mechanics, that it works against nature for man's benefit and that it is partly natural and partly mathematical, although he does not elaborate further on its scope or its relation to other arts and sciences.

Later, Moletti would also insist that mechanics is concerned not with machines as material constructions but with their properties, that is, with their capacity to multiply forces, which arises from their design.

The first extensive discussion of the scope of mechanics occurs in the commentary on the *Mechanical Problems* by another humanist and philosopher, Alessandro Piccolomini. Piccolomini (1508–1579) taught moral philosophy at Padua in 1539 and Sienna from 1546 to about 1558. In 1574 he was named Archbishop of Patras and Coadjutor to the Archbishop of Sienna.[23] Moletti described him as 'a man very skilled both in peripatetic philosophy and in the mathematical disciplines'; the sixteenth-century biographer of mathematicians Bernardino Baldi wrote that he combined 'profound peripatetic scholarship with an intense interest in mathematics,' although Baldi also expressed reservations about Piccolomini's mathematical skill.[24]

Piccolomini's works included commentaries on Aristotle's *Meteorology, Ethics,* and *Politics;* a translation of the *Poetics;* works on astronomy, natural philosophy, logic, and the certitude of mathematics; and the commentary on the *Mechanical Problems.* The last enjoyed considerable popularity; first published in Rome in 1547, it was reprinted in Venice in 1565 and translated into Italian by the metallurgist Vannoccio Biringuccio in 1582.[25] Most later commentators were familiar with Piccolomini's commentary, and Moletti seems to have followed it fairly closely, particularly on the scope and subject of mechanics and on its relation to other sciences. In the notes to the edition of the *Dialogue* that follows, I have quoted a few parallel passages from Piccolomini and cited a number of others.

In the prologue to his commentary – which was actually an extended paraphrase of the original – Piccolomini gave an extensive account of the nature and scope of mechanics and its place within the classification of the sciences. Using the traditional medieval Aristotelian division of the sciences, he established that mechanics is contemplative (speculative) rather than operative (practical), mathematical rather than natural, and equivalent in status to optics and astronomy. Philosophy, he begins, has two parts, the contemplative and the operative. The operative he divides into the active – comprising the moral, domestic, and civic – and the factive – comprising the manual or sellularian arts. Contemplative philosophy, in turn, is made up of the divine (metaphysics), the natural, which treats all moving things, and mathematics, which treats quantity apart from material. Corresponding to the two species of quantity, magnitude and number, there are two mathematical sciences, geometry and

arithmetic. And on each of these in turn depend other sciences: on arithmetic depends music and on geometry depend stereometry, *perspectiva* (optics), cosmography, astronomy, and mechanics.[26] Although these dependent sciences are not purely mathematical, since they concern matter in some way, they are more aptly called mathematical than natural, since they use mathematical instruments and proofs.[27] For Piccolomini, then, mechanics is a contemplative, mathematical science analogous to astronomy and optics and thus far removed in this classification from the practical or sellularian arts.

While not to be identified with the sellularian arts, Piccolomini asserts, mechanics nevertheless provides their causes and principles. For although the *mechanicus*, acting in the capacity of a craftsman, could design mechanical instruments and machines to do work, as a *mechanicus* he rests content with considering their causes and principles. And although mechanics imitates and aids nature, it must often work outside of or against the opposition of nature in order to accomplish human ends. Mechanics, then, is the art by which we discover machines and instruments by which the sellularian arts can overcome this opposition.[28]

If mechanics is not to be set among the sellularian arts, neither is it to be identified with natural philosophy. Piccolomini argues that although both mechanics and natural philosophy are contemplative sciences dealing with natural matter, that is, with mobile and heavy things ('mobilia' and 'ponderosa'), they differ in that mechanics treats this subject in a mathematical way, just as *perspectiva* and music do, but as natural philosophy in contrast does not. It is precisely this mathematical habit, Piccolomini asserts on several occasions, that Aristotle himself adopted in book 6 of the *Physics*, when discussing how motions can be compared.[29] The elaboration of this mathematical habit applied to motion would ultimately lead to Galileo's new mathematical science of motion, published in the Third Day of his *Two New Sciences* of 1638. But even then, the science of motion would not be considered part of mechanics. For Piccolomini, as for most writers on the *Mechanical Problems* as well as for Galileo, mechanics concerned primarily violent, not natural, motions, while natural philosophy and the science of motion derived from it concern primarily natural motions.[30]

Piccolomini, then, considered mechanics a contemplative, mathematical science, under geometry, that provides the mechanical arts with their principles and causes and that is concerned primarily with motions and effects produced against nature for the benefit of man. Moletti relied on his commentary mainly for the discussion of the scope,

subject, and place of mechanics among the other sciences. Like Leonico, however, Piccolomini brought to his commentary on the *Mechanical Problems* little mathematical expertise and little knowledge of the other mechanical traditions of antiquity and the Middle Ages. For these we have to turn to other, more mathematical, sources.

Perhaps the leading mathematician of the sixteenth century was the abbot Francesco Maurolico of Messina. That Moletti was among his pupils there would suggest that Maurolico's influence might loom large in his conception of mechanics. Surprisingly, this is not the case. Maurolico's conviction that the principles of Aristotelian mechanics are to be found in the works of Archimedes is directly contrary to Moletti's notion that Archimedean mechanics (about which he was very ill informed) depended on Aristotelian, which in turn depended on Euclid. Nevertheless, it is worth sketching Maurolico's approach to mechanics, if only to contrast it with Moletti's.

Francesco Maurolico (1494–1575) both taught as a private tutor and lectured publicly on mathematics in Messina, where Moletti was his pupil. In 1569 he was appointed professor of mathematics at the Jesuit university, and he also held various civic posts in Messina, including being in charge of the city's fortifications.[31] His mathematical works were numerous: he wrote on astronomy, optics, and sundials, as well as editing, translating, and commenting on many ancient mathematical authors, including Euclid and Archimedes. He also wrote a commentary on the pseudo-Aristotelian *Mechanical Problems*. During his lifetime, however, few of these works were published. Although its dedication is dated 1569, his commentary on the *Mechanical Problems* was not printed until 1613, well after his death. That Moletti knew this commentary is unlikely, since he had already left Messina by the date of its dedication and he never mentions Maurolico in his several lists of mechanical authors. Further, Maurolico expressly condemned those mechanics who ignore centres of gravity and are exclusively concerned with the principle of circular movement – a precise characterization of Moletti's approach.

Maurolico begins this commentary with an extensive discussion of the subject and scope of mechanics and of its place in the classification of the sciences.[32] His classification is essentially the same as Piccolomini's: he places the sellularian arts under the active branch of operative philosophy and divides contemplative philosophy into *physica, mathematica*, and *metaphysica*. Metaphysics, he notes, considers both natures that are separated from matter, such as God and Intelligences, and that which is most common to all existing things (i.e., being). *Physica* consid-

ers the principles of bodies both simple and mixed (i.e., composed of various elements) and their motions. Finally, mathematics treats quantity and form apart from the material; it has two parts – geometry, which treats continuous quantity, and arithmetic, which treats discrete quantity. When geometry and arithmetic are applied to some particular material, he continues, they generate further sciences, which are in some way intermediate ('mediae') between the mathematical and the natural. As examples Maurolico lists music, astronomy, *perspectiva*, the science of weights, stereometry, cosmography, geography, architecture, painting, sculpture, and all mechanical theory ('omnis ratio mechanica'). He notes that since these sciences derive from mathematical speculation – even though in them it is applied to particular, natural things – they should be called mathematical rather than natural. Thus mechanics, Maurolico concludes, is to be placed under the mathematical part of philosophy.

Mechanics in turn is applied to various mechanical devices ('machinamenta'). Maurolico notes that Archimedes excelled in this, although earlier Aristotle treated many such applications most acutely in his *Mechanical Problems*. But this work is marred, according to Maurolico, by poor translation and a faulty text and has thus remained obscure in spite of the efforts of many scholars, including Leonico and Piccolomini. A further difficulty, Maurolico asserts, is that in order to understand many of its problems, which treat the theory of the lever, balance, centres of gravity, and such, one must be familiar with the theory of equal static moments ('doctrina aequalium momentorum') rather than only with the properties of the circle. The doctrine of equal static moments, then, which is contained in Archimedes' *On Plane Equilibrium*, should serve as an introduction to the *Mechanical Problems*, and Maurolico criticizes previous expositors of the work for not providing this introduction (as I mentioned, this criticism would later apply to Moletti). He also criticizes the commentators for being strangers to theoretical knowledge ('a necessaria speculatione alieni') and consequently for relying too much on material and sensible instruments and experiments, just as 'simple mechanicals' ('puri mechanici'), in so far as they lack all theory, customarily do. By 'simple mechanicals' Maurolico here apparently means the practical designers and builders of machines. Maurolico proceeds to mend this defect by briefly listing several statical postulates and by defining centre of gravity and the various parts of the balance and lever. He then gives the law of the balance in terms of static moment ('momentum'). That he means static moment by this term becomes clear as he distinguishes it from the other quantities treated in mechanics: volume

('corpus'), weight ('pondus'), and a fourth, which he calls a power ('vis') or impetus. This fourth, he asserts, belongs only to moving bodies and serves to explain how a smaller weight thrown from a distance can counterbalance or even raise a larger weight that it otherwise could not. All such matters, he concludes, follow from and are demonstrated through the theory of weights ('doctrina ponderum'); they labour in vain who try to explain the proofs in the *Mechanical Problems* through sensible experiments, which was attempted only before the doctrine of equilibrium ('doctrina aequeponderantium') became known. Archimedes' accomplishment, Maurolico adds, is not diminished even if one assumes that he knew the *Mechanical Problems*, for even then he 'reworked all into a most orderly demonstration in the way of geometers' ('omnia in demonstrationem ordinatissimum, Geometrarum more, redegisset').

Although Maurolico preferred to ground his mechanics in Archimedean statics, he nevertheless recognized the application of mechanics beyond conditions of equilibrium, to the motion of heavy bodies – the moving of weights with machines and the harnessing of impetus. His notion of the primacy of Archimedean statics in mechanics came to be shared by many mathematicians who followed, including Guidobaldo del Monte, Bernardino Baldi, and Galileo, but notably not by his erstwhile pupil Moletti.

Of greater moment to Moletti was the autodidact Niccolò Tartaglia (1499 or 1500–1557), whose writings on mechanics were perhaps the model and part of the occasion for the *Dialogue*. Although lacking a formal education, Tartaglia held various teaching positions throughout his career.[33] From 1516 or 1518 he taught the abacus (elementary calculation) at Verona; from 1534 on he was professor of mathematics at Venice (teaching a year at Brescia, 1548–9). His original works include the *Nova scientia* (1537), on ballistics and other mainly military applications of mathematics, and the *Quesiti et inventioni diverse* (1546), on various mathematical and mechanical topics, including artillery, fortification, and statics.

Tartaglia played an important role in the transmission of ancient and medieval mechanical works. In 1543 he published the first vernacular translation (in Italian) of Euclid's *Elements*, accompanied by his own commentary, and in the same year he published the medieval Latin translations of several of Archimedes' works, including *On Plane Equilibrium*. In 1551 he published an Italian translation of the first book of Archimedes' *On Floating Bodies*.[34] But in spite of this familiarity with the works of Archimedes, the medieval science of weights was the main

influence on Tartaglia's mechanics. Jordanus de Nemore's *Liber de ponderibus* had been printed in 1533, and in 1539 Tartaglia acquired a manuscript of the more important *De ratione ponderis*, which Troianus Curtius printed for the first time in 1565, from materials left by Tartaglia at his death.[35]

Although Tartaglia did not write a full commentary on the *Mechanical Problems*, he devoted book 7 of his *Quesiti et inventioni diverse* to a discussion of its contents and book 8 to the science of weights, claiming that the *Mechanical Problems* could only be explained through that science.[36] Although the *Mechanical Problems* used both physical and mathematical arguments, Tartaglia wrote, its weakness lay in its appeal to physical arguments ('argomenti naturali') that could be contradicted, sometimes by other physical arguments and sometimes by mathematical arguments through the science of weights. For Tartaglia, physical arguments were those based only on empirical observation, while mathematical arguments provided all of the abstract reasoning found in mechanics. Tartaglia could thus assert the superiority of mathematical arguments over physical arguments and then explain their different conclusions as the result of material hindrances.[37]

By the 'science of weights,' then, Tartaglia meant theoretical, mathematical mechanics. Employing the scholastic terminology, he refers to that science as a subalternated science ('scientia subalternata'), noting that part of it derives from geometry and part from natural philosophy, since some of its conclusions are demonstrated ('se demostrano') geometrically and some are verified ('se approvano') physically or naturally (i.e., empirically). The two results ('costrutti') to be drawn from this science, he continues, are the ability to calculate the strength ('virtù') and power ('potentia') of any machine used to augment the strength of men; and the ability to find out how to augment this strength to any degree by means of machines.[38] These results are apparently not to be realized within the science of weights itself, however, for Tartaglia limits the rest of his discussion to definitions, axioms, and propositions concerning the balance and inclined plane, deriving most from the medieval science of weights.[39] Presumably, the results are to be realized in the application of the science of weights to problems such as those found in the *Mechanical Problems*. The science of weights itself provides only the first principles of this application.

Tartaglia's use of the term 'science of weights' to designate theoretical mechanics indicates the extent of his reliance on the medieval, Jordanean tradition. Although familiar with Archimedes' purely statical

proof of the law of the lever found in *On Plane Equilibrium*, he preferred the Aristotelian-Jordanean proof involving velocities and displacements. In presenting this proof, Tartaglia notes that Archimedes had proved the same thing, but from different principles and arguments, which 'would not be suitable in this treatise, it being of somewhat different subject.'[40] Tartaglia thus could not see his way clear to combining Archimedean statics, which was founded on the principle of centres of gravity and developed purely statically, with the Aristotelian-Jordanean approach, which used velocities. Perhaps he found the latter approach more promising for his science of weights, which treats the principles of augmenting power.

In any case, Moletti did not follow Tartaglia in finding the principles of Aristotle's mechanics in the science of weights (a tradition with which Moletti shows little familiarity), but in his *Dialogue* he did adopt a form of question and answer similar to that of Tartaglia's *Quesiti*. Further, Moletti begins the *Dialogue* by relating a conversation overheard among some engineers, and the questions he reports echo several of the questions treated by Tartaglia. The *Quesiti*, then, was possibly what Moletti had in mind when he began the *Dialogue*.

One more possible influence remains to be considered before we turn to Moletti himself. Moletti's predecessor in the chair of mathematics at the University of Padua was Pietro Catena (d. 1577), who was apparently the first to lecture on mechanics at the university. Significantly, the text on which these lectures were based was the *Mechanical Problems*. Guidobaldo del Monte attended Catena's lectures on the *Mechanical Problems* in 1564 and was not favourably impressed; nor was Bernardino Baldi, who heard them in 1573. Unfortunately, we have no way of confirming their opinion, for Catena's lecture notes apparently have not survived. We do have several other works by him, however, including one on the certitude of mathematics and another, the *Universa loca mathematica* (1556), on the mathematical passages in Aristotle's logical works. In this latter work, while discussing the *Posterior Analytics*, Catena treats the subalternation of sciences and the *scientiae mediae*, among which he includes mechanics.[41] He also cites problem 11 of the *Mechanical Problems*, which asks why it is easier to move heavy weights on rollers than in wheeled carts. Aristotle's explanation is that the wheel of a cart encounters friction at the axle, while, of course, a roller does not. Catena characterizes this explanation as physical, offering in its place the 'truly demonstrative reason' ('ratio propter quid'), that a roller moves more easily because, being smaller than a wheel, it makes a larger angle with

the surface and thus comes less in contact with it. In support of this he offers a geometrical proof he derives from some theorems of Euclid; the proof depends on being able to compare the angle between the tangent and the circumference of circles of different diameters – the so-called angle of contingence or horn angle. Catena remarks that he will describe the rollers currently in use 'in quaestionibus mechanicis.'[42] Perhaps this meant that he intended to extend the *Universa loca* beyond Aristotle's logical works to include the *Mechanical Problems*, or that he intended to write a separate commentary on it, though no mechanical work by him is known to be extant. Nevertheless, this one problem reveals several points in common between Catena and Moletti. In proposition VII of the *Dialogue*, Moletti similarly asserts that the angle of contingence is smaller the larger the circle. And, more generally, Moletti also looks to Euclidean geometry to explain mechanical effects. On the other hand, I have not found any mention of Catena in Moletti's mechanical writings, so it remains uncertain whether Moletti knew his lectures on mechanics or his other works. Perhaps the only influence that Catena had on Moletti was to die in 1576, leaving the chair of mathematics at Padua vacant. The following year Moletti left Mantua to assume this chair, and never apparently returned to his unfinished *Dialogue*.

II Life and Works of Giuseppe Moletti

Giuseppe Moletti was born in 1531 in Messina, Sicily, where he studied medicine and mathematics, the latter as a pupil of the mathematician Francesco Maurolico.[43] After spending some time in Naples and Verona, Moletti moved to Venice in about 1556, where he made the acquaintance of the Paduan bibliophile and polymath Gian Vincenzo Pinelli, who became a lifelong friend and correspondent. According to the funeral oration by Antonio Riccoboni, Moletti was esteemed in Venice as a physician ('medicus'), a great philosopher ('philosophus summus'), and an excellent mathematician ('mathematicus eximius').[44] Perhaps he taught these subjects privately or practised medicine. In 1561 he published the *Discourse on the Terms and Rules of Geography*, which was appended to a translation of Ptolemy's *Geography* and was later reprinted several times; a few years later he published an ephemerides for the years 1563–80.[45] In May of 1570, while he was still in Venice, Moletti's health was grave enough that he wrote his last will and testament.[46] Shortly afterwards he must have recovered, for he left Venice to take the position of mathematics tutor to Vincenzo, the son of Guglielmo Gonzaga, Duke of Mantua, a

position he held for seven years. Sometime probably in the early 1570s he wrote the *Discourse on How to Study Mathematics*, a sort of professional manifesto in which he sketched in Italian the main mathematical disciplines, the authors that wrote on them, both ancient and modern, and a plan of study recommended especially for princes. The section devoted to mechanics, which makes up almost one-third of the work, is of particular interest here, for it contains in an earlier form much of the material that would later be incorporated into the First Day of the *Dialogue*.[47] In my introduction to the *Dialogue* below, and especially in my notes to the translation, I have noted the main parallels. In 1576, towards the end of this period at Mantua, he began – but never finished – the work that is edited and translated here, the *Dialogue on Mechanics*.

By May of 1577 Moletti had left Mantua to take up the chair of mathematics at the University of Padua left vacant by the death of Catena in 1576. But despite his departure, Moletti maintained his attachment to the ducal court, returning to Mantua several times in the early 1580s to put himself at the service of Duke Guglielmo. By 1582 Moletti was receiving a pension from the duke of two hundred *scudi*, and in his letter of thanks he addresses him as 'my patron.'[48] In this final stage of his career he became involved in one of the great mathematical controversies of the day, the reform of the calendar, publishing his contributions in 1580.[49]

At Padua, according to the University Rolls, Moletti's teaching consisted mainly of geometry, astronomy, and optics, and his surviving lecture notes bear this out.[50] But he also lectured on mechanics, the pseudo-Aristotelian *Mechanical Problems* serving as the authoritative text. He dated the first sheet of his set of introductory lectures, titled in the manuscript *Exposition of Aristotle's Book of Mechanics*, on 6 October 1581, adding that the lectures were repeated 10 February 1582; the Rolls list him as lecturing on mechanics again in 1585–6.[51] He was apparently only the second ever to lecture at the university on the *Mechanical Problems* – Pietro Catena preceded him, and Galileo was to follow suit. Written in his own hand and in places considerably rewritten and revised, Moletti's *Exposition of Aristotle's Book of Mechanics* comprises fourteen topics that Moletti felt it necessary to discuss before dealing with the problems themselves. These topics include the title, scope, and authorship of the treatise; the nature and division of mechanics; other writers on the subject; and the utility of mechanics. The title of the lectures and several internal references suggest that the notes represent only an introduction and that Moletti would go on to discuss in a 'disorderly and

extemporaneous fashion' each of the problems in turn.[52] In the discussion of the *Dialogue* that follows and in the notes to the translation, I have drawn on these lecture notes to fill out Moletti's views on a number of points.

In this last phase of his life Moletti seems to have been plagued with poor health. The lecture notes on the *Mechanical Problems* are accompanied by a preface in which Moletti apologizes to his students for his debility.[53] He died on 26 March 1588, at the age of fifty-seven, though Riccoboni wrote that he looked seventy or eighty years old.[54]

On his death, Moletti's papers were acquired by his friend Gian Vincenzo Pinelli. What survive of them, along with the rest of Pinelli's library that escaped predation, sale, and disaster at sea, are now to be found in the Ambrosiana Library in Milan.[55] These volumes contain the farrago of commentaries, lectures, treatises, notes, and letters that I mentioned at the beginning of this introduction, including the *Dialogue on Mechanics*, which we are now ready to consider in detail.

III The *Dialogue on Mechanics*

Bound in the same volume as his lecture notes on the *Mechanical Problems* is a copy of Moletti's unfinished *Dialogue on Mechanics*.[56] This copy – not in Moletti's hand – is entitled simply 'Things all incomplete and without certain resolution' ('Cose tutte imperfette et senza risoluzione certa'), although Pinelli later added the misleading title 'Some notes on the matter of artillery' ('Alcune memorie in materia d' artiglieria') as well as the ascription to Moletti. The first page is dated 1 October 1576, that is, in Moletti's last year at Mantua; the *Dialogue* fills some twenty folios until it breaks off mid-page without a conclusion.

In addition to this, the fullest copy (= S), there are two fragments. The first, also in the Ambrosiana, is a short fragment of the *Dialogue* – about four folios' worth, but this time in Moletti's own hand.[57] This fragment (= D), which begins mid-sentence and ends mid-sentence and mid-page, has only its first few lines in common with the fuller copy and then goes off on a digression concerning the division of the sciences. It would seem that this fragment was a part of Moletti's original draft, a part that was discarded in the course of writing but was somehow preserved while the rest of the draft was lost after being copied.

The second fragment, consisting only of the last section on falling bodies and now in the Biblioteca Nazionale in Florence, was copied from S by Giambatista Venturi, who printed it in 1818 and sent his hand-

written copy to the Grand Duke of Tuscany, where it was incorporated into the collection of Galileo materials in Florence. From this copy Raffaello Caverni printed the two passages in his *Storia del Metodo Sperimentale in Italia* (1895).[58] Finally, part of Caverni's transcription was translated into English by Thomas Settle in an article in 1983.[59] As far as I know, this is the entire publishing history of the *Dialogue.*

The form of the *Dialogue* deserves some attention before I turn to its contents, since its form reveals something of Moletti's purpose. Composed in Italian rather than in Latin, the work presents a fictional dialogue taking place over two days between two interlocutors identified only by initials. In the part of the *Dialogue* transcribed by Venturi and thence known by Caverni and Favaro, the interlocutors are identified as P and A, where P offers extensive answers to the questions asked by A. Since P addresses A usually as 'Your Lordship' ('Vostra Signoria') or ('Vossignoria') but A addresses P as 'Your Highness' ('Vostra Altezza'), Caverni concluded that P stands for 'Prince' ('Principe') and signifies Vincenzo Gonzaga, Moletti's young pupil, while A stands for 'Author' ('Autore') and signifies Moletti himself. To explain why the young prince was instructing his tutor, Caverni suggested that by putting his novel doctrines in the mouth of royalty, Moletti sought to avoid the anger of the peripatetics. Favaro, however, suggested rather more plausibly that Moletti meant it as a tribute to the young prince's mathematical precosity.[60] But there is one difficulty with these identifications. In most of the rest the *Dialogue*, which was not examined by either Caverni or Favaro, the interlocutors are signified by the initials AN and PR. Now, PR could in fact stand for *Prince*. In the first sentence of the *Dialogue*, AN addresses PR as 'Your Highness' and refers to 'so many machines and instruments' he sees belonging to him. The *Dialogue* is set in Mantua – AN begins the Second Day by mentioning his stroll last evening outside the Palazzo Te, the sumptuous ducal retreat on the outskirts of the city. Towards the end of the *Dialogue*, AN mentions a certain 'Franceschino, who plays the organ of Santa Barbara'; Santa Barbara is the ducal church in Mantua and Franceschino was the usual name of Francesco di Rovigo, who was organist there in the 1570s.[61] All this suggests that Moletti set the *Dialogue* in the present in Mantua; PR, standing for *Principe*, probably meant Duke Guglielmo Gonzaga, Moletti's patron, rather than Vincenzo, who was only fourteen years old at the time. Moletti would thus be lending an air of nobility and usefulness to the study of mechanics for the edification of his pupil and at the same time flattering the mathematical knowledge of his patron.

The identification of AN is more difficult, since obviously 'AN' cannot stand for 'Autore.' There are, however, some clues within the *Dialogue* itself. AN makes reference to occurences of his own boyhood in Sicily, where Moletti himself was born, and he mentions the fortifications of Naples, where Moletti spent some time before coming to Venice. This might suggest that AN was meant to be Moletti himself. But in one passage PR suggests that AN is a practically minded man and calls him a 'cavaliero soldato.' Moletti, however, was a philosopher, mathematician, and physician, not a soldier or a knight. So it seems unlikely that AN refers to Moletti. More likely Moletti meant AN to be some real or fictional person who was representative of the audience he had in mind for his *Dialogue*, a practical man of affairs with good common sense, along the lines of Galileo's Sagredo.[62]

The *Dialogue* seems to fulfil the promise Moletti made in the *Discourse on How to Study Mathematics* that he would compose a fuller treatment of mechanics.[63] Part of his concern was to make mechanics noble and worthy to be studied by a prince, and accessible and palatable to princely tastes: thus the dialogue form and the use of Italian rather than Latin. But Moletti also had a larger purpose in mind. He sought to establish mechanics not only as a science worthy of the study of princes and useful to them, but to found mechanics on Euclid's geometry, to lay down the Euclidean principles of Aristotelian mechanics. And having done that, he sought to extend mechanics into problems concerning the relations of mover to resistance and the motion of falling bodies, problems traditionally treated in natural philosophy. To see how he did this, we must now turn to the *Dialogue* itself.

The First Day

The *Dialogue* is divided into two days. The First Day is largely taken up by a discussion of the pseudo-Aristotelian principle of circular motion as the fundamental principle of mechanics, followed by two sets of geometrical propositions offered as proof of this principle. Moletti begins the *Dialogue* with an introductory discussion that sets the tone of the entire work. AN – the practical layman – relates to the Prince a conversation he overheard among some engineers and architects ('ingegneri et architetti') concerning various mechanical effects. In particular, one engineer asked another why a smaller cannonball can sometimes go farther than a larger one but make less impact; and why heavy weights can be lifted with a block and tackle and how to calculate the resulting mul-

tiplication of force. The answers given had something to do with circles and centres that AN did not fully understand, so he appeals to PR for an explanation. As I suggested above, this reported conversation may allude to Tartaglia's treatment of similar topics in the form of questions and answers in his *Quesiti*. Further, Tartaglia offered precisely the sort of explanation of the impact of cannonballs that has confused AN: he analysed their paths into curved and straight parts, and then reduced their different effects to the balance and in turn to the principle of circular motion.[64]

Prompted by these practical instances, the Prince names mechanics, the art of the engineer, as that science that provides the principles and causes of such effects. Citing and following closely the pseudo-Aristotelian *Mechanical Problems*, the Prince then defines mechanics as the science by which we discover how to overcome by art difficulties imposed by nature; for example, how to move a great weight with a small power, or how to prevail with a smaller number of soldiers over a larger number. With this last Moletti is appealing to the ancient, broader notion of mechanics that included any sort of machination or tactic that gives one an advantage, though for him the connection to the narrower notion of mechanics is direct. Archimedes, as Moletti never tires of reminding us, defended Syracuse against a superior Roman force by means of his ingenious machines. The military advantage of overcoming a large military force with a smaller is thus subsumed by the general advantage conferred by machines of overcoming any larger force with a smaller.[65] In the *Discourse on How to Study Mathematics*, Moletti gave an extensive discussion of how mechanics (rather than architecture, as Vitruvius had asserted) serves the civil science ('scienza civile') by helping to defend cities against external enemies or against those who seek to oppress the communal liberty, but he cautions that this does not mean that the mechanic is to usurp the command of armies.[66] Nor is mechanics to be confused with the so-called mechanical arts – metalworking, carpentry, and the like. For these the Prince prefers the word 'sellularian,' since they are characterized by actual execution and labour. The true mechanic works only with his mind, not with his hands, in designing and planning the work, while craftsmen are his instruments for executing it. For mechanics, the Prince claims, is a theoretical and mathematical science subalternated to geometry. In the earlier draft of the *Dialogue* Moletti then veered off into a digression on the division of the sciences, with special attention to Aristotle's notion of the middle sciences, which included mechanics and astronomy, and how they differ from pure

mathematics on the one hand and natural philosophy on the other. In trying to show how astronomy and natural philosophy can differ despite their treating the same subject matter, he tried to develop an analogy to ironworking: how making armour and making hoes differ despite their both using iron as their material. But before he could bring this analogy to bear on astronomy, he apparently decided he was digressing and pulled what he had written. I have edited and translated this digression in the Appendix below.

This same topic – the division of the sciences and the relation of the middle sciences to geometry on the one hand and natural philosophy on the other – was taken up by Moletti in several other works, notably in the earlier *Discourse on How to Study Mathematics* and again in his later *Commentary on the Mechanical Problems*. In the latter Moletti would give an elaborate account of the division of the sciences and of mathematical abstraction in order to identify mechanics as a contemplative, mathematical science subalternated to geometry, an account that is quite similar to Piccolomini's and Maurolico's.[67] He goes on, however, to analyse the relation between mechanics and geometry in terms of the medieval theory of subalternation. One science is subalternated to another, he asserts, when each considers the same subject but in a different way, in that the subalternated science adds an accidental difference to this subject. Optics, for example, adds the physical quality visible to line, the subject of geometry; where the geometer makes a geometrical line from a physical line by abstraction, the student of optics makes from this geometrical line a physical line by addition. In the case of mechanics, the added accident, according to Moletti, is circular motion. For like other intermediate sciences (and Moletti uses the term 'scientiae mediae'), namely, harmonics, optics, and astronomy, mechanics applies mathematical arguments and demonstrations to sensible things and concerns sensible matter and motion, whereas pure mathematics concerns only abstract quantity. It is this purely abstract quantity that the intermediate sciences 'contract' or apply to sensible matter, making what Moletti calls 'sensible quantity' ('quantum sensibile'). The subject of mechanics, Moletti concludes, is not simply machines, but rather sensible quantity mobile in circular motion, or machines whose form or principle is circular motion or compounded from it. The properties of mechanics are the powers and virtues of such machines for lifting and drawing weights and for throwing projectiles.[68]

In the *Dialogue*, however, Moletti is content to have the Prince assert that the subject of mechanics is the machines used and the effects they

produce through their properties; the goal of mechanics is to explain such properties and effects through their principles and causes. And this is exactly what Moletti then proceeds in general to do. Following the *Mechanical Problems*, the Prince argues that the properties of machines can be reduced back to certain marvellous properties of the circle, in particular that a point on its radius moves faster the farther it is from the centre. This in turn gives rise to the central principle of Aristotle's mechanics, that the same power is the swifter and more effective the farther from the centre it is applied. In this First Day, Moletti seems to take it as self-evident that the swifter motion will cause the greater force, and that the greater force will cause the swifter movement; only in the Second Day does he try to analyse the relation between speed and force.

The principle of mechanics is then proved in two different ways with two sets of geometrical propositions. Although the propositions are presented as part of the fictional dialogue, they are clearly meant to be formal and rigorous demonstrations. They are accompanied by carefully drawn and lettered compass-and-ruler diagrams and they refer repeatedly to Euclid's *Elements*. On the other hand, they are not presented in a rigorously deductive order; rather, the main proposition – the principle of circular motion – is given first and supported by various suppositions, and only afterwards are these suppositions themselves demonstrated. In the manuscript the propositions are not set off from the general discussion in any way. In order to lay bare the structure of Moletti's mathematical argument (such as it is) and for ease of reference, I have numbered the propositions and given each a formal enunciation.

First, to illustrate the principle of circular motion – that the farther a point is from the centre of a circle, the faster it is moved and the more effective it is – the Prince generally follows the discussion in the pseudo-Aristotelian *Mechanical Problems*. Interestingly, in the earlier *Discourse on How to Study Mathematics* Moletti followed roughly the order of the pseudo-Aristotelian text, reproducing right down to the lettering the diagram illustrating the principle of circular movement that accompanied Leonico's translation.[69] In the *Dialogue*, however, he has adopted his own order: he gives the principle first, simplifying the diagram and lettering it differently, and only afterwards does he prove the supporting propositions. This suggests that when he came to write the *Dialogue* Moletti had moved beyond the *Mechanical Problems* to reflect more independently on the principle of mechanics and how to establish it.

The Prince begins by making two suppositions, both of which are found in the *Mechanical Problems*: first, that a circle is described by a line

moving with two motions, one natural and the other against nature, and, second, that these motions are devoid of any ratio to each other in any times; for when two motions are in some fixed ratio, they will describe a straight line. (These two suppositions he will later prove as proposition V and proposition II.) Now, he continues, if a line were to be moved with only one motion (call this its natural motion), it would describe a rectangle (see figure I); but if one end of the line were fixed, each point of the line would be deflected into a circle, and the closer a point is to the centre, the more deflected from a straight line it would be and the smaller the circle it would make. This deflection towards the centre, being opposed to the natural motion, the Prince calls violent, and since the closer point is more deflected from its natural motion than the farther, it is more constrained and so will be less effective. Therefore, the farther from the centre a power acts, the more effective it will be, which is the pseudo-Aristotelian principle of circular movement.

Note that this principle does not simply assert the obvious fact that a point farther from the centre of a rotating radius is moved more swiftly. Rather, it asserts that circular motion is caused by the combination of two motions – a natural, rectilinear motion together with an opposing violent motion caused by the restraint of the immobile centre – and that a smaller circle results from a greater constraint.

Now, to confirm this principle means, for Moletti, to demonstrate the geometrical propositions on which it depends. According to Aristotle, a science cannot prove its own principles; in the case of a middle science like mechanics, its principles are proved in a higher science, here geometry. So while the principle of circular movement is mechanical – it deals geometrically with powers and motions, both natural and violent – the propositions that support it are geometrical. What I have called proposition I, then, sets out to prove that the chord in a smaller circle is farther from the circumference than an equal chord in a larger circle. In effect, this proposition is geometrical proof that for an equal vertical (or natural) movement, a point moving on a smaller circle is more deflected horizontally (or violently). The Prince claims that although this can be proved in several ways, he will offer the easiest that he has yet found. First, to illustrate the geometrical proposition, he places an equal chord in each of two unequal and concentric circles, and notes that we are to prove that the distance between the chord and the circumference will be greater in the smaller circle (see figure II). But instead of proceeding with this diagram, he redraws the two circles, this time with the smaller inside and tangent to the larger, and then, citing *Elements* 1.33 and 34, he

draws the chord in the smaller circle and again in the larger such that they are perpendicular to the diameter drawn through the point where the circles touch (see figure III). Now, since the smaller circle is inside the larger, its chord will be below that of the larger. And since the chord in the smaller circle is below the chord in the larger, its perpendicular distance from their point of contact is longer, which is what was to be proven. The leap in this proof, however, is in this last step – that the chord in the smaller circle is 'below' the parallel and equal chord in the larger – which is unproved and which amounts to a *petitio principii.* Moletti does cite *Elements* 3.14 here – that equal chords in a circle are equally distant from the centre – though he does not cite 3.15 – that the chord nearer the centre of a circle is longer – which would seem more relevant to his purpose. Note, however, that even 3.15 concerns only unequal chords in the same circle, while Moletti's proposition concerns equal chords in unequal circles. At best, then, Moletti has given a bare sketch of a possible proof, and he seems to rely more on inspection and intuition than rigorous geometric demonstration.

The next four propositions (propositions II–V) return to prove the suppositions made earlier in stating the principle of mechanics and to prove several other related propositions. Proposition II demonstrates that two rectilinear movements in any fixed ratio will describe a straight line, which comes directly out of the *Mechanical Problems.*[70] Interestingly, at the end of the proof, the Prince calls AN's attention to an instrument he has on hand that mechanically produces this effect, presumably one of those many mentioned at the beginning of the *Dialogue.*

The next propositions, which are not discussed in the *Mechanical Problems,* concern the combinations of various movements and the resulting path. Proposition III proves that two circular movements combined can describe a straight line. This is now called the Tûsî-couple, named for the medieval Islamic astronomer al-Tûsî, though the Prince credits the proposition to Copernicus in the *De revolutionibus,* and his proof is a rough paraphrase of Copernicus's. This proposition, although not at all necessary for the proof of the mechanical principle, gives the Prince the occasion to defend Copernicus by saying that his detractors have not understood his intent. What he means is that those who condemn Copernicus for holding that the earth goes around the sun have failed to understand that he meant this only as a hypothesis to save the astronomical appearances, and not as literally true.[71]

Proposition IV, similarly, is unnecessary to Moletti's purpose here, though perhaps he included it because it was original: the Prince claims

that he discovered it himself when he 'heard the *Mechanics*,' i.e., when he was taught the *Mechanical Problems*. This proposition proves that a straight motion combined with a circular motion can describe a straight line. In effect, it asserts that the path traced out by a point on a circle simultaneously rotating and moving rectilinearly can be a straight line. What is not specified is that for this to happen, both of the motions cannot be uniform: if the circle is rotating clockwise at a uniform speed, then its centre must move to the left at a speed varying as the sine of the angle. Again, at the end of the proof, the Prince points out that instruments could be made to illustrate this effect, though he does not seem to have one at hand.

Proposition V, that the two movements that produce a circle are devoid of any ratio, was mentioned earlier by the Prince in the discussion of the principle of mechanics and does appear the *Mechanical Problems*. Leonico seems to have had some doubts about this conclusion; in the commentary accompanying his translation he ventured cautiously that it is not without merit that, if two motions are in no ratio, then they might describe a circle. Piccolomini was more definite: he allowed that it might seem that two motions devoid of a ratio would not necessarily produce a circle but perhaps some other figure, but in the end he defended the conclusion.[72] As for Moletti, in his *Discourse on How to Study Mathematics* he raised two doubts concerning this passage in the *Mechanical Problems*, doubts which he claimed no one else had raised. The first doubt was exactly Piccolomini's, that two movements devoid of any ratio may not necessarily describe a circle, but some other curve; and the second, its converse, that the two movements that do describe a circle may not in fact be devoid of any ratio. In response to the first doubt he argued that only a circle can be described by two movements devoid of any ratio, for the ellipse, the hyperbola, and the parabola all are produced by two motions in some ratio to each other, which in the case of the ellipse is somehow expressed in the ratio of the major to the minor axis. He seems unaware that for a circle this ratio is one of equality. As for the second doubt, he showed that there is no fixed ratio between horizontal and vertical displacements of a circular motion.[73] In the *Dialogue*, he repeats the main arguments from the *Discourse* almost verbatim, although he omits the diagrams and illustrations. And of course there is no *fixed* ratio between the speeds of horizontal and vertical motions that compose circular motion, but neither Piccolomini nor Moletti, let alone the author of the *Mechanical Problems*, had any way to describe circular motion as an analytical function.

AN then asks whether the principle of Aristotelian mechanics can be proved in another way, and the Prince replies with a second set of propositions, deriving in part from Tartaglia and ultimately from the medieval science of weights. Instead of starting with circular motion and analysing it into its natural and violent components, this second set of propositions begins by considering the speeds of heavy bodies falling first along inclined planes and then along circular arcs, and then presents a series of geometrical propositions to support the conclusion that bodies will fall more swiftly – and thus act more effectively – along the arc of the larger circle. The first of these propositions, proposition VI, which seems to have been meant as preliminary to proposition VII, asserts that a heavy body falls most swiftly along a vertical line, and less swiftly the more the line is removed from the vertical. A similar proposition is found in the central work of the medieval science of weights, Jordanus de Nemore's *De ratione ponderis*, though Jordanus says that the body is 'heavier' rather than 'swifter' along the vertical. Tartaglia, in his treatment of the science of weights in book 8 of his *Quesiti*, accepts this proposition as evident and adopts it as a *petitio* or supposition.[74] For his part, the Prince takes as evident only that the vertical path will be swiftest, noting that this is obvious to the senses. He then goes on to show that the motion would be slower the more removed from the vertical is the plane along which the heavy body is constrained to move; if the plane were horizontal, it would not be moved at all. Now, if equal distances were marked off on inclined planes descending from a common point, a heavy body would take more time in descending that distance the farther from the vertical it is. And if there were equal resistances at the ends of those equal distances, the heavy body would make a greater impact the closer to the vertical it fell. Moletti will discuss the relation between speed, resistance, and the force of impact in greater detail in the Second Day, so I shall have more to say on it when I come to discuss that part of the *Dialogue*.

Proposition VII, the central proposition in this second set, extends the same considerations to a heavy body constrained to fall along the arc of a circle from the point where the circle is tangent to a vertical line, and concludes that the greater the circle the swifter the motion, a case that Tartaglia had included with the *petitio* mentioned above.[75] The Prince notes that this case is similar to Aristotle's principle of circular motion. The reason that the motion is swifter along the arc of the greater circle, he states, is that the larger the circle, the less its circumference is removed from a straight line. By this he seems to mean that if a body

were to fall along a circular arc from its point of tangency with the vertical, since the arc of a larger circle is closer to that vertical straight line than the arc of a smaller, then the body would fall more swiftly the larger the circle. And again, if one were to mark off an equal arc length on the various circles, bodies would traverse this length more swiftly on the arc of larger radius. This proposition seems to follow in some way from the previous one concerning inclined planes, though the Prince does not explicitly state the connection.[76]

The remaining two propositions in the First Day are proofs of the geometrical foundations of this conclusion, setting out to prove that the larger the radius of an arc, the less it is removed from a straight line. The Prince goes about this in two ways, first by arguing from the angle of contingence, and then by comparing the length of the arc subtended by an equal chord in unequal circles. The angle of contingence or 'horn angle,' as it is sometimes called, is the angle formed between the circumference of a circle and a tangent. It is thus a 'mixed' angle, one arm of which is a straight line and the other a curve, and it gives rise to a number of paradoxes when one tries to compare its size to rectilineal angles or to horn angles of different circles. Euclid, in *Elements* 3.16, proved only that the horn angle is smaller than any rectilineal angle. In giving the same argument in the *Discourse on How to Study Mathematics*, Moletti took the opportunity to rebut Jacomo Pelettario (Jacques Peletier), who in his commentary on *Elements* 3.16 contended that horn angles were not quantities at all. Moletti took the contrary view, later seconded by the great Jesuit mathematician Christopher Clavius, that horn angles were quantities that can be compared in size at least among themselves, if not with rectilinear angles.[77] In the *Dialogue*, Moletti contents himself with asserting that because the larger circle makes a smaller angle of contingence with its tangent, its circumference is less removed from a straight line and a body will fall so much the swifter along it.

He expends considerably more effort to prove that the arc subtended by a chord in a smaller circle will be longer (and thus, by implication, more removed from a straight line) than the arc subtended by the same chord in a larger circle. His method here is to construct on the chord EF (see figure IX) in the smaller circle an arc EOF equal to the arc CID subtending the equal chord in the larger circle, and proving that this arc will fall between the chord and the arc EMF of the smaller circle. He then reasons that, since arc EMF contains arc EOF (which is equal to arc CID), and the container is larger than the contained, that arc EMF is larger than arc CID. And therefore, since they are on the same chord, arc EMF

is more curved than arc CID. Again, Moletti seems to rely here more on intuition and common sense than on rigorous geometrical reasoning. The major premise, that the container is larger than the contained – even if one grants that one arc can 'contain' another – is of doubtful geometrical validity. Moreover, Moletti does not define what 'more curved' means. One could, for instance, define an arc as more curved if it represents a greater proportion of the whole circumference of its circle than another, equal arc, though Moletti does not do so. His proof is thus plausible but not demonstrative.

From this he draws as a corollary the last proposition in the First Day, that when there are equal arcs in unequal circles, the arc in the larger circle subtends a longer chord than the one in the smaller. This he proves by first drawing equal chords in unequal circles, and then noting that by the previous proposition the arc subtended in the smaller circle will be longer than that in the larger (see figure X). He then marks out on the circumference of the larger circle an arc equal in length to the arc of the smaller circle, and shows that the chord it subtends is longer than that subtended by an equal arc in the smaller circle. This corollary is not needed for Moletti's argument, and seems to have been added out of exuberance.

All of these propositions, concludes the Prince, are intended to demonstrate the geometrical foundations of the fundamental principle of Aristotelian mechanics, that a weight or force constrained to move in a circle will be that much more effective the farther it is from the centre. In this way Moletti has tried to establish geometrically the principle of mechanics and so realize in fact the subalternation of mechanics to geometry.

While the principle of circular movement is the fundamental principle of Aristotelian mechanics, Moletti also admits at the end of the First Day several other principles into mechanics, though they seem of much lesser importance to him here. The first of these is the vacuum, to which he attributes the effects of water machines. Mines (i.e., explosives) and artillery, in contrast, depend on the multiplication of air. And still other machines, such as things that cut (i.e., saws and knives), fortification, ordnance, and ships, depend on their various shapes for their effectiveness. He suggests that each of these has its own, proper principles, although he insists that for those that can be reduced to the lever, the principle of circular motion is 'almost universal.' He gives only the merest mention of it here, but the vacuum (or the 'spirit' or the 'air' as he also calls it) is, together with the circle, one of the two main principles of

mechanics in his *Discourse on How to Study Mathematics.* There he reduced all mechanical effects, including the war machines of antiquity, artillery, and pumps, to one or other of these principles, or to both.[78] In the *Dialogue,* the Prince then gives a brief division of mechanics into wheeled (i.e., geared) machines, water machines, ships, fortification, artillery (including mines and artificial fire), and deployment of troops and artillery. He notes that the science of weights, which includes Archimedes' *On Plane Equilibrium,* is under mechanics, and he expresses the hope to discuss it another day. The First Day ends with the Prince promising to show AN the next day how the principle of circular movement can be used to solve mechanical problems.

The Second Day

If it was Moletti's original intention in the Second Day to apply the principle of mechanics to the solution of particular problems, he never quite got around to it, though he began well enough. The Second Day, like the First, opens with a casual discussion of practical, everyday things. AN mentions that as he was strolling last evening outside the Palazzo Te, he came upon some boys who were throwing reeds with the aid of spear-throwers, and he noticed that when two of his footmen tried to throw them with great force, the reeds merely vibrated violently and fell only a short distance away. This, he remarks, must have something to do with mechanics.[79] The Prince agrees, promising to explain it later; but first he will use the principle of circular movement to explain the archetype of all mechanical movement – why heavy weights can be moved by a lever. In the movement of the lever, and in mechanical movement in general, he begins by distinguishing two 'principal things' – the weight to be moved and the power needed to move it. Beyond these, he recognizes two 'additional things' – the lever or instrument itself and the fulcrum or, in general, the 'immobile.' Much of the Second Day is then devoted to discussing the proper ratio between the moving power and the resistance, and the necessity that there be something immobile from which the motion arises.

The idea that every motion must arise from something immobile Moletti attributes to Aristotle's *De motu animalium,* a translation of which Leonico had conveniently printed in the same volume as his translation of the pseudo-Aristotelian *Mechanical Problems.*[80] The immobile principle seems to serve Moletti as the link between mechanics and natural philosophy. In mechanics, the centre of circular movement or the

fulcrum of the lever is the immobile principle; the Prince quotes the famous saying attributed to Archimedes, that if he were given an (immobile) place to stand, he could move the world. And in natural philosophy, according to Aristotle in the *De motu animalium*, the immobile principle appears in the motions of both animate and inanimate bodies. In the case of animate bodies, one can distinguish external immobiles (e.g., the ground over which a man walks) from internal immobiles (e.g., the shoulder when one moves one's arm). Ultimately all motion is grounded in the universal immobile, the unmoved or prime mover.[81] The Prince then illustrates for AN the external and internal immobile principles for various sorts of motions, including the movement of snakes, the motion of a boat pushed by the wind, and swimming. In all cases of animate motion, there is an instrinsic immobile as well as an extrinsic one. The Prince notes that the immobile principle is of great moment in mechanics.

Moletti intended to treat the qualities and properties of each of the principal and additional things in full, but the *Dialogue* breaks off while he is still discussing the moving power and the resistance. With these topics Moletti is now firmly on ground traditionally covered in natural philosophy: the ratio between the mover and the moved; the role of resistance and of the medium; the relative speeds of different weights in free fall; and the cause of natural acceleration. Let me sketch briefly what he has to say on each of these.

First, the Prince asserts that moving powers can be either natural, that is, instrinsic to the thing moved, or violent, that is, extrinsic and applied from without. Mechanics, he notes, concerns principally – but not exclusively – violent, extrinsic powers. Now, for movement to arise, the moving power cannot be either less than or equal to the thing moved – the Prince quotes the peripatetic tag 'action does not arise from the equal' ('ab equali non provenit actio'). So it must be greater. But greater by more than an 'insensible ratio' – we would say 'more than infinitesimally greater.' This means that Moletti was not prepared to recognize, as Galileo did later, that the addition of only an infinitesimal weight could disturb the equilibrium of the balance. But neither can the moving power exceed the moved body by a very large ratio. To explain why, the Prince turns to the role of resistance in motion, making the provocative statement that resistance is a cause of motion. While this may appear paradoxical, the Prince insists he will not indulge in vain quibbling over words as he might in talking with some superannuated philosophers; rather, with a *cavaliero soldato* such as AN, he will reason with principles

that are manifest to sense. It then becomes apparent that by resistance he means here not the external resistance of the medium, but rather the internal resistance of the mobile itself. This internal resistance is the 'cause' of motion in the sense that a very light body, i.e., one with very little internal resistance to motion, once set in motion almost immediately comes to rest. But a heavy object, although initially resisting the motion, once moving will continue for a great distance. Thus a very great force is lost on a very light object, for very little motion results. AN remarks that Aristotle, in the *Mechanical Problems*, treated a similar question, and the Prince identifies it as question 34.[82] And so in general, the moving force must be proportioned to the body to be moved, being neither overly large nor overly small. This general principle suggests an explanation for the example that opened the Second Day: why a reed when launched with excessive force goes only a short distance. The Prince also uses it to explain why the impact of a cannonball is greater the greater the resistance it encounters. For if it meets with little resistance, it can do little damage no matter how fast it moves; but if it encounters great resistance, it can do great damage.[83] Curiously, Moletti touches only in passing on the cause of the continued motion of projectiles, mentioning very briefly the opinion of some (including Aristotle) that the air is the cause, and of others, who attribute it to the moving power impressing itself on the projectile, an allusion to medieval impetus theory.[84]

We now come to that part of the *Dialogue* transcribed by Venturi and printed by Caverni, where the Prince asserts that different weights of different materials fall from the same height simultaneously, and where he suggests that he has done the experiment many times himself.[85] He knows of course that Aristotle had apparently suggested quite the opposite, that heavy bodies fall at speeds in proportion to their weights. At one time he thought he had found a way to 'save' Aristotle on this point, but now thinks better of it. AN persuades him to explain it nonetheless. In essence, it consists of identifying the speed of fall with the force of impact, so that Aristotle's statement would mean that heavy bodies fall with 'speeds,' i.e., produce impacts, in proportion to their weights. The Prince's dissatisfaction with this interpretation lies in the fact that a lighter body, falling from a great height and with great speed (in the usual sense), will often make less impact than a heavy body falling from much less height and with much less speed. This dramatizes the two incompatible definitions of speed, and Moletti is apparently not willing to abandon the usual one in order to save Aristotle.

The Prince then offers Girolamo Cardano's explanation why two unequal weights should fall simultaneously, but again he is less than satisfied with it. Cardano had argued that the power of descending is proportional to the weight, that a falling body compresses the medium beneath it in proportion to this weight, and that the resistance offered by the compressed medium is in proportion to its density. Thus a ball three times the diameter of another of the same material and thus weighing twenty-seven times as much has twenty-seven times the power of descending. But it also compresses the medium to twenty-seven times the density, and thus encounters twenty-seven times the resistance. Its greater power of descending, then, is exactly balanced by the greater resistance it encounters, so it descends at the same speed as the smaller ball. Now, the Prince is willing to allow that the medium is compressed, but he is not satisfied that it will be compressed in exactly the ratio necessary, an assumption that Cardano does not prove.[86] Had Moletti applied his notion of internal resistance to falling bodies, so that the heavier body would have exactly that much more *internal* resistance to motion as it has more power of descending, he would have come very close to the concept of inertial free fall. But while internal resistance works well when applied to violent motions, where the moving power is extrinsic to the thing moved, if applied to natural motions it produces a paradox. For how can one and the same weight be both the cause of a body's motion and at the same time offer the resistance?

Finally, the Prince touches on the vexed question of the cause of the acceleration of natural motions.[87] He lists what he calls the three most famous explanations. (1) By place, whereby a heavy body in seeking its natural place goes faster the closer it gets to it. This opinion was commonly cited by medieval authors, including Jean Buridan, who attributed its origins to Averroes. Moletti dismisses this explanation on the grounds that it implies that a heavy body must have some sort of cognitive power by which it knows where it is, which he considers unreasonable.[88] (2) By movement, whereby movement, being the act or perfection of a mobile thing, is more easily acquired the more movement the body already has, so the faster and faster it moves. AN suggests here an analogy to the exercise of a moral virtue or of a skill like playing the organ: the more one does it the easier it gets.[89] And (3) by the medium, the explanation that Moletti seems to favour. In an argument attributed to Cardano but which I have been unable to find in his works, the Prince argues that as a body falls, the column of medium beneath it that must be compressed becomes shorter, and thus the faster the body can go. To

this explanation the Prince makes what seems to be his own addition: that if the body were to start its fall from the sphere of the moon, there would be no air behind it to fill the vacuum it leaves, so that it would fall very slowly at first. But once it has fallen some way, there would then be air to rush in behind it, which would help to speed its fall. This additional cause was of course Aristotle's explanation of the continued motion of projectiles. At this point the *Dialogue* breaks off, so we are left not knowing whether this was Moletti's last word on the question.

IV Mechanics after Moletti

In the *Dialogue on Mechanics* Moletti has attempted two remarkable goals. In the First Day, he sought to establish the geometrical foundations of Aristotelian mechanics – to articulate the geometical principles that underlie mechanics and thus to realize in fact the subalternation of mechanics to geometry. The importance he attaches to the pseudo-Aristotelian *Mechanical Problems* is not unusual, for among sixteenth-century writers on mechanics it was considered one of the chief works in the ancient mechanical tradition, though most other commentators supplemented it with Archimedes, Hero, Pappus, or Jordanus.[90] Moletti in fact borrowed at least indirectly from Jordanus through Tartaglia, but although he mentioned the names of Hero and Pappus, he apparently was not familiar with their works. More surprising is his complete neglect of Archimedean statics and the concept of centres of gravity; despite his boundless praise for Archimedes' mechanical abilities, he seems completely unfamiliar with his works. While mathematicians such as Maurolico, Commandino, and Guidobaldo saw Archimedes as applying rigorous mathematics to Aristotle's vague physical notions, Tartaglia had explicitly rejected Archimedes' statical approach to the law of the lever in favour of the dynamic approach of Jordanus and the medieval science of weights, relegating the *Mechanical Problems* to a position subordinate to the science of weights.[91] For Moletti, however, Archimedean statics was subalternated to Aristotelian mechanics because Archimedes took as his principle, without proof, what Aristotle had proven naturally. For Aristotle had given the physical cause of the longer arm of an equally weighted balance descending, while Archimedes accepted without proof that it does descend.[92] Moletti's own contribution was to lay the Euclidean foundations for Aristotle's mechanics.

In the Second Day, as preliminary to applying the mechanical princi-

ple of circular movement to specific mechanical problems – and this was his second remarkable goal – he tried to determine the rules that govern the quantitative relation between moving powers and the resistances, and to bring under mechanical scrutiny various problems of natural and violent motion that traditionally were treated mainly in natural philosophy. What linked mechanics and natural philosophy for Moletti was the immobile principle as discussed by Aristotle in the *De motu animalium*: in the case of mechanical movement the immobile was the centre of rotation or fulcrum; in the case of natural motion, it was the instrinsic or extrinsic immobile described by Aristotle, and ultimately the unmoved mover, the source of all motion in the world. Yet despite this link, Moletti was not prepared to identify mechanics and natural philosophy; rather he saw them as distinct sciences that had to some extent a common subject matter. For a fuller account of Moletti's views of the relation between mechanics and natural philosophy, we must turn to his later *Commentary on the Mechanical Problems*.

In discussing there whether mechanics is an art or a science, Moletti argued that sciences differ from arts both in their subjects and in their ends: the subjects of sciences are necessary and eternal, while those of arts are subject to our will; and the end of sciences is knowledge of causes and the truth, while the end of arts is productive work. And because the principles and causes of machines are necessary and eternal and in no way subject to our will, mechanics is a science not an art.[93] The principles and causes of mechanics are thus almost what we should call 'laws of nature'; Moletti in fact asserted that nature herself makes use of mechanics. For mechanics is found in all the works of nature, although the difference between our use of mechanics and hers, he added, is that nature always knows the means to achieve her ends, while we often do not know the means to achieve ours.[94] He conceded that in books 6, 7, and 8 of the *Physics* Aristotle does deal with problems that involve some mechanics, but notes that there Aristotle treats them in a natural rather than in a mechanical way. According to Moletti, Aristotle's conclusions concerning such problems, although entirely within the bounds of natural philosophy, can subsequently be assumed as principles or properties in mechanics.[95] Mechanics and natural philosophy are thus distinct sciences, just as mechanics and geometry are.

Moletti's mechanics, like Tartaglia's, was fundamentally dynamic, concerned with powers overcoming resistances to produce motions and useful work. For this reason it is not surprising that problems of natural and violent motion should arise in it. Such problems – the ratio of mov-

ing power to resistance, the speed and acceleration of falling bodies, the motion and effects of projectiles, and the like – were usually treated within the confines of natural philosophy, which was concerned with nature and motion in general. That Moletti took on such problems was also perhaps at least partly because sixteenth-century mathematicians tended to take on problems involving military technology, especially fortification and ballistics, often treating the motion of cannonballs and their force of impact along with the more traditional mechanical devices. Tartaglia, for instance, first in 1537 and again in 1546, had written extensively on ballistics and fortification; Francesco Maurolico, writing sometime before 1569, had included *vis* or *impetus* among the mechanical quantities; and Galileo, in a mechanical treatise written around 1593, included the force of impact as one of the five simple machines.[96] Artillery was arguably the most characteristic machine of the bellicose sixteenth century, a machine that worked not with gears and levers, however, but solely through the explosive power of gunpowder and the laws of natural and violent motion. Clearly, to understand and harness this machine was of great practical and political importance to a sixteenth-century Italian prince. Moletti's *Dialogue on Mechanics*, written while in the service of such a prince, opens with a discussion of the motion and effects of cannonballs.

But despite his attempt to lay the geometrical foundations of Aristotelian mechanics and to incorporate into it the rules governing the motion of heavy bodies, Moletti never saw mechanics in the modern sense, as the universal mathematical science of bodies in motion. Nor did he envisage, as Galileo would later, a new mathematical science of motion distinct both from mechanics and from natural philosophy. Moletti believed that mechanics could be established on Euclidean principles as a middle science alongside the other mathematical sciences (such as astronomy, optics, and harmonics), distinct from but sharing with natural philosophy certain principles and problems.

Moletti was not entirely successful in achieving these two goals. His attempt to Euclidize Aristotle's mechanics failed not so much because of its occasional lapses of rigour, but because of its irrelevance. The future of mechanics lay not with Euclid and Aristotle, but with Archimedes, as Maurolico had insisted long before. Though Moletti was slow to recognize this, the mathematician and biographer of mathematicians Bernardino Baldi grasped the importance of Archimedes even while commenting on the pseudo-Aristotelian *Mechanical Problems.* I have described elsewhere in some detail how Baldi tried to reconcile the two.[97]

In his 'Life of Archimedes,' one of the many mathematical biographies he composed, Baldi asserted that Aristotle, 'leaving aside the mathematical aspects, preferred to draw his demonstrations from physical principles,' which, if combined with mathematics, would result in a complete theory of mechanics. This combination, according to Baldi, was what Archimedes had accomplished. For Aristotle correctly adduced the velocity of the greater circle as the cause of a lever's moving a weight more easily, but he left the rule indeterminate. Archimedes, accepting Aristotle's principle, went on to determine the precise mathematical relation between the force and weight – that the weight is to the force as the inverse of the lengths of their respective arms.[98] In the Preface to his commentary on the *Mechanical Problems*, written in the 1580s, Baldi tried to bring the work up to date by including accounts of the five simple machines and centres of gravity. The primary principle of mechanics, he asserted, is the principle of static equilibrium ('centrobarica'), which is based on centre of gravity ('gravitatis centrum'); this principle, advanced by Archimedes in his *On Plane Equilibrium*, had been unknown to Aristotle. Baldi proposed to remedy this defect by confirming the problems in the Aristotelian work with 'mechanical, that is, Archimedean proofs.'[99]

But while Baldi was seeking to establish the Archimedean foundations of Aristotelian mechanics and Moletti the Euclidean, Guidobaldo del Monte (1545–1607) was discarding Aristotelian mechanics entirely and replacing it with Archimedean, introducing into mechanics a new method and rigour. His *Liber mechanicorum*, which was printed in Latin in 1577, just a year after the date of Moletti's *Dialogue*, and published in Italian in 1581, would become perhaps the most influential mechanical treatise of its time.[100] It represented the full application of Archimedean statics to the theory of the five simple machines for moving weights on the model provided by Pappus.[101] Guidobaldo's *Liber mechanicorum* constitutes a thorough repudiation of the dynamical approach to mechanics taken in the medieval science of weights and adopted by Tartaglia, and as found in the *Mechanical Problems*.[102] The book thus marked the end of the substantive contributions of the *Mechanical Problems* to the science of mechanics, and rendered all Moletti's work in mechanics obsolete.

Yet Moletti's ill-fated attempts to establish Aristotelian mechanics on a Euclidean foundation and to extend mechanics into the realm of natural philosophy do bear comparison to the achievements of the mathematician who succeeded him in the chair at the University of Padua and who would accomplish, though in a very different way, what he had glimpsed only vaguely.

Moletti and Galileo

In the fall of 1587, less than a year before his death, Moletti received for assessment several original propositions on centres of gravity presented by the young Galileo Galilei in his bid for the chair of mathematics at the University of Bologna. In his appraisal Moletti cautiously wrote that the lemma and theorem 'seem sound to me,' and he judged the author to be 'a good and well-trained mathematician.'[103] For despite his exuberant admiration for Archimedes as the defender of Syracuse, Moletti, as I have noted before, was almost entirely unacquainted with his works and made no use of Archimedean concepts such as centres of gravity in his own. Perhaps recognizing this defect, and just a week after signing the appraisal, Moletti began to compose annotations on the first book of Archimedes' *On the Sphere and Cylinder*, though he managed to cover only three propositions before his death a few months later.[104] For Moletti it was too late: the Archimedean revival had left him behind.

Although unsuccessful in landing the position at Bologna, and after teaching at Pisa for several years, in 1592 Galileo would assume the chair of mathematics at the University of Padua left vacant since Moletti's death in 1588. The University Rolls show that in 1598, during his tenure at Padua, he, like Catena and Moletti before him, lectured on the *Mechanical Problems*.[105] Although his notes for these lectures are now lost, several of his surviving writings from around this period show that the *Mechanical Problems* had had a strong influence on his notion of the science of mechanics.

In his earliest surviving work on mechanics, dating from 1593–4, Galileo described mechanics as the science that provides the principles for producing effects with machines.[106] And, like Guidobaldo and Baldi, he noted that all of the simple machines can be reduced to the lever and ultimately to the balance. Finally, Galileo explicitly included the force of impact among the simple machines, as Maurolico had done implicitly when he recognized that a moving body has the power to produce an effect. Here, at least, Galileo's notion of the scope of mechanics is well in line with the tradition of the commentaries on the *Mechanical Problems*. But he went far beyond them to incorporate into the mechanics of movement the central insight of Archimedean statics. In *Le meccaniche*, written about 1600, he grasped the key to transforming the static balance into the dynamic lever, which had eluded Guidobaldo and Baldi, for he realized that it takes only the addition of 'any minimal heaviness' to set a balance in motion. He also recognized the usefulness of the principle of

virtual velocities (which he credited to the *Mechanical Problems)* in analysing such conditions of equilibrium and disequilibrium.[107] This achievement represents the culmination of the program adopted by Maurolico, Guidobaldo, and Baldi of incorporating Archimedean statics into a mechanics of moving weights.

Around the same time as his earliest mechanical works, Galileo had also tried to apply Archimedean principles to the motion of falling bodies, composing a dialogue and a treatise on motion, which, together with some memoranda, are now known collectively as *De motu antiquiora* (the 'older works on motion').[108] These works – unpublished until the last century – bear special comparison to Moletti's *Dialogue on Mechanics.* Galileo's general purpose in them, like Moletti's in the Second Day of the *Dialogue,* was to explore the relation between mover and moved in natural and violent motions. And like Moletti, Galileo applied a mechanical principle to the analysis of motion, though it was Archimedes' hydrostatics rather than pseudo-Aristotle's circular movement.

Galileo's starting point, like Moletti's, was to inquire concerning the paths and speeds of cannonballs of different weights: Galileo asked why the straight part of the flight of cannonballs is longer the closer to the vertical they are shot, and why heavier balls are shot more swiftly and go farther than lighter. As we saw above, Moletti's *Dialogue* began with an inquiry concerning the range of cannonballs, and in the Second Day he considered the motive force acquired by cannonballs of various materials. Both Galileo and Moletti rejected Aristotle's theory that the medium is the cause of the continuation of projectile motion, and both argued that heavier bodies can receive a greater impressed force than lighter and thus go swifter and farther, though Moletti qualified this where the bodies are very heavy and the motive forces very small, and where the bodies are very light and the motive forces very great. Both Moletti and Galileo considered the speeds of bodies falling along inclined planes and along the arcs of circles. But where Moletti compared the speeds of bodies falling along inclined planes to motions along circular arcs of different radii in order to confirm the principle of circular motion, Galileo analysed the speeds of bodies falling along inclined planes into motions along the tangents of the circle made by the ends of a balance. Note that despite this difference of approach they both saw such motions in terms of mechanics and tried to explain the various speeds with the principle of the lever.

Galileo's popular fame has rested on his assertion that all heavy bod-

ies fall with the same speed, and on his having dramatically confirmed this by dropping things from the Leaning Tower of Pisa. But in his early writings on motion Galileo did not assert this revolutionary conclusion without qualification; there he claimed only that bodies of the same density fall with the same speeds, and he tried to derive the different motive forces, and thus different speeds of various falling bodies, from the difference between their specific gravities and that of the medium through which they fall. In general, Galileo's dynamics in the *De motu antiquiora* was based on the principle of buoyancy, which he found in Archimedes' hydrostatics, and except for the special case of very flat bodies, he tended to ignore the resistance of the medium. He also suggested that falling bodies would not continue to accelerate indefinitely, but that they would reach some natural maximum speed in the medium through which they fell. It was Moletti, rather, who stated the rule of equal speeds in free fall in the more general form – that 'a ball of wood, either larger or smaller than one of lead, let fall from the same height at the same time as the lead ball, would descend and touch the earth or ground at the same moment in time' – and he suggested that he had confirmed this result many times with actual tests. And finally, he asserted that the speed of a falling body would increase to infinity if there were an infinite space through which it might fall. But in trying to explain the equal speeds of falling bodies, Moletti confused buoyancy and resistance, and the best he could offer was the explanation suggested by Cardano: that the medium under the heavier body would be more compressed and thus offer greater resistance in exactly the ratio of its weight to the lighter, rendering their speeds equal. At least he had the good sense to doubt whether the medium would in fact be compressed at all, and, if it were, whether it would be compressed in exactly the required ratio, although he could not offer a better explanation himself. Note that Moletti was not at all motivated by the anti-Aristotelianism displayed by some of his contemporaries, notably by Benedetti and sometimes Galileo. Instead, he tried to save the usual interpretation of Aristotle, that bodies fall with speeds proportional to their weights, by suggesting that speed be measured as force of impact, though this interpretation was unsatisfactory to him. Curiously, in his first attempts at articulating the principle governing the speeds of falling bodies, Galileo similarly toyed with measuring speed as the force of impact.[109]

And finally, both Moletti and Galileo discussed the cause of the acceleration of falling bodies. Moletti, as we saw, attributed this acceleration tentatively to the decrease of resistance offered by less medium that

remained under the falling body the farther it fell; Galileo dismissed this explanation along with several others, and, in a brilliant insight, suggested instead that the acceleration was caused by the ever-decreasing motive power that originally projected the body upwards or sustained it before falling.

All these similarities, however, should in no way suggest that Galileo had read Moletti's work; in fact, there is no evidence of any direct borrowing, and it is highly unlikely that Galileo, who read few printed books, should have taken the trouble to read an unfinished manuscript by a minor predecessor, even if he had known of its existence. Rather, the similarities are attributable to a common set of problems and concerns, and a shared set of sources. They bespeak a common culture shared by mathematicians in the late sixteenth century, and the more we know of this culture from the work of mediocre men like Moletti, the better we shall be able to judge the novelty and genius of men of the first rank like Galileo.

Galileo is also renowned for eschewing the Latin of academic commentaries and treatises and writing instead dialogues in Italian aimed at a more popular audience. His early considerations on motion, though in Latin, were probably first put into the *Dialogue on Motion*, a fictional conversation between Domenico, the inquiring layman, and Alessandro, the instructor. The more complete treatise that makes up the *De motu antiquiora* was probably composed afterwards, perhaps because Galileo grew impatient with the dialogue form or frustrated with trying to write a lively dialogue in the Latin that such an academic topic seemed to demand of a university professor. Only after he left the university for the patronage of the Grand Duke of Florence, following the publication of his astonishing telescopic discoveries in 1610, did he adopt for his major works the sinuous and ironic Italian that would make his literary as well as his scientific reputation, especially in his most famous and controversial work, the *Dialogue Concerning the Two Chief World Systems* of 1632.

In this respect, Moletti's career was the inverse of Galileo's. Moletti spent his earlier years at the Gonzaga court, and the sort of works he wrote during that period, notably the *Dialogue on Mechanics*, were determined in part by the conditions of his patronage. The *Dialogue*, as I suggested above, was written in Italian and in a way to appeal to the princely tastes of his patron and to recommend the study of mechanics to princes and practical men. But when Moletti assumed the chair of mathematics at Padua, his works took on a more academic aspect: he

wrote in Latin, and mostly in the form of commentaries on teaching texts. The contrast with his earlier works is most notable in his commentary on the pseudo-Aristotelian *Mechanical Problems*, which shows little of the innovation and few of the promising insights found in the earlier *Dialogue*. If Moletti had remained at court, would he have finished the *Dialogue* and followed these up?

Despite the similarities between Moletti's *Dialogue* and Galileo's *De motu antiquiora*, there was one signal difference that would prove decisive for Galileo's later achievements: where Moletti was content with stating that there was some ratio between motive powers and resistances and the resulting speeds, Galileo tried to find out exactly what that ratio actually was and what the precise rule was that governed it. As is well known, Galileo would later abandon the hydrostatical buoyancy approach to falling bodies, as well as the search for the cause of acceleration, and establish his new science of motion on the principle that natural acceleration is the acquisition of equal speeds in equal times, a principle that he discovered through measurements made with inclined planes. His new science of motion was explicitly a new, mixed science, inspired by Archimedes and modelled after astronomy, optics, harmonics, and mechanics, but as distinct from mechanics as it was from natural philosophy.

Moletti, in contrast, never envisaged a new and distinct mathematical science of motion, but rather seems to have intuited, however vaguely, that mechanics and motion were somehow to be explained by the same principles. The *risolutione certa* that the *Dialogue* lacked – the creation of a mechanics that unified under one set of principles all motions, natural, violent, and mechanical – would not come until much later, when Isaac Newton wrote his *Principia mathematica philosophiae naturalis*. Small wonder, then, that Moletti's mechanics, like most of his other works, was left unfinished.

V The Edition and Translation

The unfinished *Dialogue on Mechanics* is extant in only one complete manuscript, Milan, Biblioteca Ambrosiana MS. S 100 sup., ff. 294r–318r (= S), which is not in Moletti's hand (see plate 1). In addition, there are two short fragments: Milan, Biblioteca Ambrosiana MS. D 235 inf., ff. 59r–62v (= D); and Florence, Biblioteca Nazionale Centrale MS. Galileo 329, ff. 3r–8v (= G).[110] The fragment in G, as I mentioned above, was

copied from S by Venturi in the early nineteenth century and sent to the Grand Duke of Tuscany in Florence, where it was incorporated with Galileo's papers.[111] It is therefore not an independent witness to the text.

The second fragment, in D, is more interesting. It consists of a gathering of two leaves (four folios or eight pages) bound in a volume of other incomplete and fragmentary works by Moletti. This fragment, which is in Moletti's characteristic scrawl (see plate 2), begins mid-sentence and has only a few lines in common with S before it goes off on a digression concerning the subject of mathematics, the nature of mathematical abstraction, and the difference between astronomy and natural philosophy, when it breaks off mid-page. My guess is that this fragment is all that remains of the original draft of the *Dialogue*: when Moletti realized that he was digressing, he stopped writing mid-page and set aside the last four folios he had written. He then went back to the beginning of the digression, copied out the first few lines of the discarded pages on a new sheet, and proceeded on with the text as we now have it. After a fair copy (now in MS. S) had been made, the original draft was presumably discarded and is now lost. But the four extracted folios survived among Moletti's papers, and were later bound into MS. D. Since the digression is in Moletti's hand and discusses matters that were of considerable interest to him, I have edited and translated it in the appendix.

The edition of the main text is therefore based almost entirely on the one manuscript witness, S, though I have used the holograph fragment in D for models of orthography, and I have collated Venturi's partial copy (= G) with S where it offered readings that differed in sense. In the apparatus I have noted only meaningful variants, ignoring differences of spelling, punctuation, and the order of words.

The orthography of MS. S, various as it sometimes is, has generally been left almost unchanged. Accents have been placed following the usage of the holograph fragment D; where accents vary on the same words in D, the more frequent variant has been imposed on the text. In S the copyist seems not to distinguish between grave and acute accents, marking them with a roughly vertical stroke or dot; in cases where S shows an accent but D offers no model or analogy, I have printed the accents as in modern Italian, with the grave on the last syllable of oxytones (e.g., *haverà, quantità)* and on a few monosyllables (e.g., *è* = is). Otherwise, apart from distinguishing the consonant *v* from the vowel *u* and printing *ij* as *ii*, I have not tried to modernize or normalize the Ital-

ian. The few differences in spelling from modern Italian are little impediment to understanding the text; archaic spellings and archaic words and usages I have listed in a brief glossary at the end of the text, giving the modern Italian equivalent or an English translation.

Moletti's Italian in this work can sometimes be quite stilted and courtly; if rendered literally into modern English, it would often ring false. In my translation I have tried to satisfy the requirements of both accuracy and the modern ear. For instance, AN addresses the Prince throughout as 'Vostra Altezza' ('Your Highness'), and the Prince in turn addresses him as 'Vostra Signoria' ('Your Lordship'), occasionally adding 'illustrissima' ('most illustrious'). After the first instance, I have rendered these forms simply as 'you.' More difficult for the translator – or for any careful reader – is the frequent looseness of Moletti's sentences. Both in the Italian edition and in my translation I have ruthlessly broken his sometimes endless periods into coherent sentences and grouped them into paragraphs. I have also numbered the separate mathematical propositions that Moletti presents in the First Day and given each an enunciation. I hope that this editorial intrusion will better reveal the logical and mathematical structure of Moletti's arguments as well as make for readier reference. In the translation I have wrestled awkward and ungrammatical passages (such as sentences that change subject midstream) into modern grammatical form, though, apart from supplying the occasional missing word within pointed brackets, I have left the Italian as the copyist wrote it.

In the notes to the English translation I have tried to include all of the sources that Moletti drew upon, both those he cited and, where I could find them, those he did not. In the first part of the First Day, Moletti quotes extensively from the pseudo-Aristotelian *Mechanical Problems*, and in the Second Day he refers to Aristotle's *De motu animalium* and the *De incessu animalium*; in the notes I have given these quotations in the Latin translations of Leonico, which Moletti had at hand. The many parallels to Moletti's earlier *Discourse on How to Study Mathematics* and his later *Commentary on the Mechanical Problems*, some of them almost word for word, are also quoted at length, since neither of these works has yet been printed. Where Moletti cites propositions from Euclid in the course of a proof, I have given the enunciations in Sir Thomas Heath's translation; I have not yet determined what version of the *Elements* Moletti himself used. I have also tried to explain the allusions he makes to Aristotelian philosophy, most of which were common knowledge then but today are known only by specialists. Where appropriate I have included

references to recent secondary literature, though usually a fuller discussion and bibliography will be found in the Introduction.

The figures are redrawn from the neat compass-and-ruler figures of MS. S (see plate 2). Occasionally I have had to add lettering (within <>) where it was missing on the diagram but called for in the text.

NOTES TO THE INTRODUCTION

1 *Discorso universale ... nel quale son raccolti, & dichiarati tutti i termini, & tutte le regole appartenenti alla Geografia*, printed with *La Geografia di Claudio Tolomeo*, tr. Girolamo Ruscelli (Venice, 1561); ed. Gio. Malombra (Venice, 1564), and printed once separately (Venice, 1573); *L'Efemeridi di M. Gioseppe Moleto Matematico. Per anni XVIII* [i.e., 1563–80] (Venice, 1563), extended to 1584 in the following year (Venice, 1564); *Tabulae Gregorianae motuum octavae sphaerae* (Venice, 1580); *De corrigendo Ecclesiastico Calendario libri duo* (Venice, 1580). On Moletti's printed works, see Antonio Favaro, 'Amici e corrispondenti di Galileo Galilei: XL. Giuseppe Moletti,' *Atti del Reale Istituto Veneto di Scienze, Lettere ed Arti* 77 (1917–18), 45–118, at 51–75; rpt. in *Amici et Corrispondenti di Galileo*, 3 vols., ed. Paolo Galluzzi (Florence: Salimbeni, 1983), 3:1585–1656, at 1589–1613 (henceforth cited only in the original edition, since its page numbers are preserved in the reprint).

2 Paolo Revelli has shown that the *Discorso ... della grandezza de' Stati, Domini de cinque frà li più potenti Regi e Signori dell'Universo* (1590) attributed to Orazio Malaguzzi is substantially identical to Moletti's 'Discorso che il Re Catolico sia il maggior principe del mondo,' Milan, Biblioteca Ambrosiana MS. P 145 sup., ff. 32r–41v ('Un trattato geografico-politico de Giuseppe Moleti: "Discorso che il Re Cattolico sia il maggior principe del mondo" [1580–81],' *Aevum* 3 [1927]: 417–54, at 425).

3 Moletti's extant papers are in Milan, Biblioteca Ambrosiana; the works mentioned are in MSS. D 235 inf., D 442 inf., S 100 sup., and S 103 sup.; most of his letters are also in the Ambrosiana, notably in MSS. A 71 inf., D 332 inf., S 80 sup., and S 105 sup. Some of Moletti's works in the Ambrosiana are listed in Adolfo Rivolta, *Catalogo dei codici Pinelliani dell' Ambrosiana* (Milan: Tipografia Pontificia Archivescovile S. Giuseppe, 1933), 192, 247–53 (note that the codex listed on 253 as 'P 442 inf.' should be 'D 442 inf.'). Adriano Carugo has compiled a more complete list of Moletti's works contained in the Ambrosiana in 'L'Insegnamento della matematica all' Università di Padova prima e dopo Galileo,' in *Storia della Cultura Veneta*, ed. Girolamo Arnaldi and Manlio Pastore Stocchi, 4 vols. in 7 (Venice: Neri Pozza Editore, 1976–86), 4/II (1984): 151–99; Moletti is discussed on 170–85, and the catalogue of his works is on 178–80.

4 For a fuller account of the scope and status of mechanics, see W.R. Laird, 'The Scope of Renaissance Mechanics,' *Osiris*, 2nd Series 2 (1986): 43–68, which I have followed here.

5 On medieval divisions of the sciences in general, see James A. Weisheipl,

'Classifications of the Sciences in Medieval Thought,' *Mediaeval Studies* 27 (1965): 54–90. For a recent introduction to medieval dynamics and kinematics see John E. Murdoch and Edith D. Sylla, 'The Science of Motion,' in *Science in the Middle Ages*, ed. David C. Lindberg (Chicago: Univ. Chicago Press, 1978), 206–64. For the different contexts in which these topics were treated, see, e.g., the selections in Marshall Clagett, ed., *The Science of Mechanics in the Middle Ages* (Madison: Univ. Wisconsin Press, 1959); see also John E. Murdoch, 'Philosophy and the Enterprise of Science in the Later Middle Ages,' in *The Interaction between Science and Philosophy*, ed. Yehuda Elkana (Atlantic Highlands, NJ: Humanities Press, 1974), 51–74; and Murdoch, 'From Social to Intellectual Factors: An Aspect of the Unitary Character of Late Medieval Learning,' in *The Cultural Context of Medieval Learning*, ed. John E. Murdoch and Edith D. Sylla (Dordrecht: Reidel, 1975), 271–339.

6 See Ernest A. Moody and Marshall Clagett, eds., *The Medieval Science of Weights* (Madison: Univ. Wisconsin Press, 1952); Joseph E. Brown, 'The *Scientia de Ponderibus* in the Later Middle Ages' (PhD diss. Univ. Wisconsin 1967), and Brown, 'The Science of Weights,' in *Science in the Middle Ages*, ed. Lindberg, 179–205. On the medieval tradition of Archimedes see Marshall Clagett, *Archimedes in the Middle Ages*, 5 vols., vol. 1 (Madison: Univ. Wisconsin Press, 1964); vols. 2–5 (Philadelphia: Memoirs of the American Philosophical Society, 117, 125, 137, 157) (1976, 1978, 1980, 1984). Motion, in the form of the principle of virtual displacements, was fundamental to the science of weights; see for example Pierre Duhem, *Les origines de la statique*, 2 vols. (Paris: A. Hermann, 1905–6), 2:291–301; and Moody and Clagett, *The Medieval Science of Weights*, 5–6. Part 4 of Jordanus's *De ratione ponderis* contains problems involving the motions of bodies through resisting media (see Moody and Clagett, 213–19).

7 On views of the mechanical arts in the Middle Ages see Guy H. Allard, 'Les arts mécaniques aux yeux de l'idéologie médiévale,' in *Les arts mécaniques au moyen âge*, ed. G.H. Allard and S. Lusignan (Montreal: Bellarmine; Paris: Vrin, 1982), 9–31; George Ovitt, Jr, 'The Status of the Mechanical Arts in Medieval Classifications of Learning,' *Viator* 14 (1983): 89–105; and Elspeth Whitney, 'Paradise Restored: The Mechanical Arts from Antiquity through the Thirteenth Century,' *Transactions of the American Philosophical Society* 80, part 1 (1990). On the transmission in the Middle Ages of technical knowledge see Bert S. Hall, 'Production et diffusion de certains traits de techniques au moyen âge,' in *Les arts mécaniques au moyen âge*, 147–70.

8 See, e.g., Albertus Magnus, *Posteriora analytica* 1.3.7, in *Opera omnia*, ed. August Borgnet, 38 vols. (Paris: Louis Vivès, 1890–9), 2: col. 85 a.

9 See Paul Moraux, *Les listes anciennes des ouvrages d'Aristote* (Louvain: Editions

Universitaires, 1951), 20; T.L. Heath, *A History of Greek Mathematics*, 2 vols. (Oxford: Clarendon Press, 1921), 1: 344–5.

10 The parallels between Jordanus's approach and that of the *Mechanical Problems* have often been pointed out; see for example Moody and Clagett, *The Medieval Science of Weights*, 11–14; on some possible direct influences, see Marshall Clagett, 'Three Notes: The *Mechanical problems* of Pseudo-Aristotle in the Middle Ages, Further Light on Dating the *De curvis superficiebus Archimenidis*, Oresme and Archimedes,' *Isis* 48 (1957): 182–3, at 182.

11 Paul Lawrence Rose and Stillman Drake, 'The Pseudo-Aristotelian *Questions of Mechanics* in Renaissance Culture,' *Studies in the Renaissance* 18 (1971): 65–104; see also M. Schramm, 'The Mechanical Problems of the *Corpus Aristotelicum*, the *Elementa Iordani super Demonstrationem Ponderum*, and the Mechanics of the Sixteenth Century,' in *Atti del Primo Convegno Internazionale di Ricognizione delle Fonti per la Storia della Scienza Italiana: I Secoli XIV–XVI*, ed. Carlo Maccagni (Florence: Barbèra, 1967), 151–63.

12 Pseudo-Aristotle, *Mechanical Problems*, 847a11–24, ed. and tr. W. S. Hett, in *Aristotle, Minor Works* (Loeb Classical Library) (London: Heinemann; Cambridge, Mass.: Harvard Univ. Press, 1936), 330–1; my paraphrase differs on several points from Hett's translation. The Greek text has been edited most recently by Maria Elisabetta Bottecchia, in *Aristotle, Mechanika*, Studia Aristotelica 10 (Padua: Antenore, 1982).

13 Pseudo-Aristotle, *Mechanical Problems*, 847a24–9, ed. and tr. Hett, 330–1.

14 Aristotle, *Physics*, 2.2, 193b23; *Metaphysics*, M.3, 1078a16; *Posterior Analytics*, 1.9, 76a24; 1.13, 78b37.

15 On Aristotle's treatment of these sciences see Hippocrates George Apostle, *Aristotle's Philosophy of Mathematics* (Chicago: Univ. Chicago Press, 1952), 131–9; Martin F. Reidy, 'Aristotle's Doctrines Concerning Applied Mathematics' (PhD diss., Univ. Toronto, 1968); and Richard D. McKirahan, Jr, 'Aristotle's Subordinate Sciences,' *British Journal for the History of Science* 11 (1978): 197–220. The key passages are translated and discussed in Sir Thomas L. Heath, *Mathematics in Aristotle* (Oxford: Clarendon Press, 1949), 11–12, 44–7, 58–61, 98–100. For an account of the medieval tradition see Steven J. Livesey, '*Metabasis:* The Interrelationship of the Sciences in Antiquity and the Middle Ages' (PhD diss., Univ. California, Los Angeles, 1982); and W.R. Laird, 'The *Scientiae Mediae* in Medieval Commentaries on Aristotle's *Posterior Analytics*' (PhD diss., Univ. Toronto, 1983).

16 Pseudo-Aristotle, *Mechanical Problems*, 847b16–848a15, ed. and tr. Hett, 332–5.

17 A convenient list of the thirty-five problems can be found in Rose and Drake, 'The Pseudo-Aristotelian *Questions of Mechanics*,' 71–2; several of the prob-

lems are discussed in A.G. Drachmann, *The Mechanical Technology of Greek and Roman Antiquity* (Copenhagen: Munksgaard, 1963), 13–18.

18 See Laird, 'Scope of Mechanics.'

19 Vittore Fausto, *Aristotelis Mechanica Vittoris Fausti industria in pristinum habitum restituta ac latinitate donata* (Paris, 1517); on Fausto see Rose and Drake, 'The Pseudo-Aristotelian *Questions of Mechanics*,' 77–8; and Paul Lawrence Rose, *The Italian Renaissance of Mathematics* (Geneva: Droz, 1975), 11.

20 Niccolò Leonico Tomeo, *Aristotelis Quaestiones mechanicae*, in *Opuscula nuper in lucem aedita* (Venice, 1525); Moletti mentions Fausto several times in his *In librum Mechanicorum Aristolelis expositio*, Milan, Biblioteca Ambrosiana MS. S 100 sup., ff. 156r, 157r, 158v.

21 'Aristotelicos libros graeco sermone Patavii primus omnium docuit,' Charles H. Lohr, 'Renaissance Latin Aristotle Commentaries: Authors So–Z,' *Renaissance Quarterly* 35 (1982): 164–256, at 193; and Lohr, *Latin Aristotle Commentaries II. Renaissance Authors* (Florence: Olschki, 1988), 453; 'Lingue graece valde peritus, et peripatetice philosophiae in hoc patavino gimnasio professor egregius' (Moletti, *Expositio*, f. 158v). On Leonico's career and works see also Rose and Drake, 'The Pseudo-Aristotelian *Questions of Mechanics*,' 79–80; D. De Bellis, 'Nicolò Leonico Tomeo, interprete di Aristotele naturalista,' *Physis* 17 (1975): 71–93; and D.J. Geanakoplos, 'The Career of the Little-Known Renaissance Greek Scholar Nicholas Leonicus Thomaeus,' *Byzantina* 13 (1985): 355–72. Moletti mentions Leonico's translation in his *Expositio*, ff. 157r, 158v, and quotes from it verbatim (but without citation) in his *Discorso sulla matematica* (Milan, Biblioteca Ambrosiana MS. S 103 sup., ff. 159v and 167v).

22 Leonico, *Aristotelis Quaestiones mechanicae*, ff. xxiiir–v.

23 On Piccolomini's life and works see Charles H. Lohr, 'Renaissance Latin Aristotle Commentaries: Authors Pi–Sm,' *Renaissance Quarterly* 33 (1980): 623–734, at 624–6, and Lohr, *Latin Aristotle Commentaries II*, 329–30. Note that many of Piccolomini's works, including most of his commentaries on Aristotle, were written in Italian.

24 'Vir tum in peripatetica philosophia, tum enim in mathematicis disciplinis valde versatus' (Moletti, *Expositio*, f. 157r); and E. Narducci, 'Vite Inedite di Matematici Italiani scritte da Bernardino Baldi,' *Bullettino di Bibliografia e di Storia delle Scienze Matematiche e Fisiche* 19 (1886): 625–33 – quoted is Rose's paraphrase in *The Italian Renaissance of Mathematics*, 261. For Baldi's reservations concerning Piccolomini as a mathematician, see Rose, 266.

25 Rose and Drake, 'The Pseudo-Aristotelian *Questions of Mechanics*,' 82. On Piccolomini's works, in addition to Lohr cited above, see Rufus Suter, 'The Scientific Work of Alessandro Piccolomini,' *Isis* 60 (1969): 210–22; and

Giulio Cesare Giacobbe, 'Il *Commentarium de certitudine mathematicarum disciplinarum* di Alessandro Piccolomini,' *Physis* 14 (1972): 162–93.

26 Stereometry is mentioned here separately from geometry ultimately because in one passage Aristotle put mechanics under stereometry rather than under geometry (*Posterior Analytics*, 1.13, 78b37–8). Several medieval commentators, including Aegidius Romanus and Paulus Venetus, subsequently identifying mechanics with the mechanical arts, put mechanics under stereometry and stereometry under geometry, leaving mechanics at two removes from geometry (see Laird, 'The *Scientiae Mediae*,' 145, 200); other commentators, including Themistius, Albertus Magnus, Thomas Aquinas, and Jacopo Zabarella, claimed that stereometry was simply a part of geometry (see Laird, 82, 110–11, 229–30).

27 Alessandro Piccolomini, *In Mechanicas quaestiones Aristotelis paraphrasis* (2nd ed., Venice, 1565), ff. 3r–5r; for medieval opinions on whether such sciences were more mathematical or more natural, see Laird, 'The *Scientiae Mediae*,' 6–7, 27–31, 77, 105–6, 137–8, 194.

28 Piccolomini, *In Mechanicas quaestiones*, ff. 5v, 7v–8r.

29 Piccolomini, *In Mechanicas quaestiones*, ff. 8r–v; and (on *Physics* 6) 4v–5r, 22r, and Piccolomini, *De certitudine mathematicarum* (printed with *In Mechanicas quaestiones)*, f. 107r–v. For a modern view see James Jope, 'Subordinate Demonstrative Science in the Sixth Book of Aristotle's *Physics*,' *Classical Quarterly* 22 (1972): 279–92.

30 See Piccolomini's discussion of question 32 on projectile motion, where he dismisses natural motion as irrelevant: *In Mechanicas quaestiones*, ff. 55r–56r (56 is misnumbered as 64).

31 On Maurolico's life and works see Francis J. Carmody, 'Autolycus,' in *Catalogus translationum et commentariorum: Medieval and Renaissance Latin Translations and Commentaries*, ed. Paul Oskar Kristeller, F. Edward Cranz, and Virginia Brown, 7 vols. (Washington: Catholic Univ. of America Press, 1960–92), 1:167–8; Arnaldo Masotti, 'Maurolico, Francesco,' in the *Dictionary of Scientific Biography* (hereafter *DSB)*, 16 vols. (New York: Scribner, 1970–80), 9: 190–4; Rose, *The Italian Renaissance of Mathematics*, 159–84; Lohr, 'Renaissance Latin Aristotle Commentaries: Authors L–M,' *Renaissance Quarterly* 31 (1978): 532–603, at 573–4, and Lohr, *Latin Aristotle Commenataries II*, 252; Marshall Clagett, 'The Works of Francesco Maurolico,' *Physis* 16 (1974): 149–98; Clagett, *Archimedes in the Middle Ages*, 3:749–70; and the articles by Edward Rosen listed by Clagett on 750n of the last reference. In his *Discorso sulla matematica*, Moletti asserts that Maurolico was *il 'mio precettore,'* f. 147v.

32 Francesco Maurolico, *Problemata mechanica cum appendice*, ed. Silvestro Mau-

rolico (Messina, 1613), 7–10; the entire preface, along with several of the questions, is printed in Clagett, *Archimedes in the Middle Ages*, 3:784n–787n.

33 On Tartaglia's life and works see Arnaldo Masotti, 'Tartaglia, Niccolò,' in *DSB*, 13:258–62; Stillman Drake and I.E. Drabkin, *Mechanics in Sixteenth-Century Italy* (Madison: Univ. Wisconsin Press, 1969), 16–26; and Rose, *Italian Renaissance of Mathematics*, 151–4.

34 See Clagett, *Archimedes in the Middle Ages*, 3:538–607.

35 The fate of the Jordanus tradition in the sixteenth century has yet to be fully written; for Tartaglia's role see, in addition to the works cited in note 33 above, Duhem, *Origines de la statique*, 1:194–208.

36 Niccolò Tartaglia, *Quesiti et inventioni diverse* (Venice, 1554; facs. ed. Brescia: Ateneo di Brescia, 1959); parts of book 7 and all of book 8 are translated in Drake and Drabkin, *Mechanics in Sixteenth-Century Italy*, 104–43; see William A. Wallace, *Galileo and His Sources: The Heritage of the Collego Romano in Galileo's Science* (Princeton: Princeton Univ. Press, 1984), 203–5.

37 Tartaglia, *Quesiti*, fols. 78r–v; tr. Drake and Drabkin, *Mechanics in Sixteenth-Century Italy*, 105–7. Another who took this unusual view of the relative roles of the physical and the mathematical in mechanics was the French commentator Henri de Monantheuil at the end of the century: see Henri de Monantheuil, *Mechanica Graeca, emendata, Latina facta, et Commentariis illustrata* (Paris, 1599), 12–13; on Monantheuil see Rose and Drake, 'The Pseudo-Aristotelian *Questions of Mechanics*,' 99–100; and Lohr, 'Renaissance Latin Aristotle Commentaries: Authors L-M,' 593–4, and, *Latin Aristotle Commentaries II*, 269.

38 Tartaglia, *Quesiti*, f. 82v; tr. Drake and Drabkin, *Mechanics in Sixteenth-Century Italy*, 111.

39 Drake and Drabkin have conveniently noted these parallels in the footnotes to their translation in *Mechanics in Sixteenth-Century Italy.*

40 Tartaglia, *Quesiti*, 93v; tr. Drake and Drabkin, *Mechanics in Sixteenth-Century Italy*, 134.

41 Pietro Catena, *Universa loca in logicam Aristotelis in mathematicas disciplinas* (Venice, 1556), 65–6, 81–3; on Catena see Favaro, 'I Lettori di Matematiche,' 63–4; Rose and Drake, 'The Pseudo-Aristotelian *Questions of Mechanics*,' 93; Rose, 'Professors of Mathematics at Padua,' 302, and *Italian Renaissance of Mathematics*, 222, 243, 286; Charles H. Lohr, 'Renaissance Latin Aristotle Commentaries: Authors C,' *Renaissance Quarterly* 28 (1975): 689–741, at 707, and *Latin Aristotle Commentaries II*, 86; and Giulio Cesare Giacobbe, 'La riflessione metamatematica di Pietro Catena,' *Physis* 15 (1973): 178–96.

42 Catena, *Universa loca*, 80–1.

43 Some of the material in this and the following sections appeared in my article

'Giuseppe Moletti's "Dialogue on Mechanics" (1576),' *Renaissance Quarterly* 40 (1987): 209–23. The latest treatment of Moletti's life and works, with special emphasis on his philosophy of mathematics, is Carugo, 'L'Insegnamento della matematica,' 170–85, and 'Giuseppe Moleto: Mathematics and the Aristotelian Theory of Science at Padua in the Second Half of the 16th Century,' in *Aristotelismo Veneto e Scienza Moderna*, ed. Luigi Olivieri (Padua: Antenore, 1983), 1:509–17. See also Favaro, 'Amici e corrispondenti di Galileo Galilei: XL. Giuseppe Moletti'; Favaro, 'Giuseppe Moletti,' in A. Mieli, ed., *Gli Scienziati Italiani* (Rome, 1921), 1.1:36–9; Favaro, 'I lettori di matematica nella Università di Padova,' *Memorie e Documenti per la Storia dell' Università di Padova* 1 (1922): 3–70, at 67–70; Favaro, *Galileo Galilei e lo Studio di Padova* (Padua: Antenore, 1966), 104–5; Paul Lawrence Rose, 'Professors of Mathematics at Padua University 1521–1588,' *Physis* 17 (1975): 300–4, at 303; and Rose, *The Italian Renaissance of Mathematics*, 286–7 *et passim*. The main sources for Moletti's biography are the funeral oration by Antonio Riccoboni, *Orationum volumen secundum* (Padua, 1591), ff. 41v–46v; and Aloysius Lollinus, *Vite: Josephus Moletius messanensis*, Belluno, Biblioteca Civica MS. 505. cart. s. XVII, ff. 75v–77v. Moletti's name took several forms in the sixteenth century: in Italian, the usual form was 'Gioseppe Moleto,' which was how Moletti signed his letters and the title pages of his Italian works; in Latin the usual form was 'Josephus (or Iosephus) Moletius.' Modern scholars are divided on how the name should now be spelt: Caverni and Carugo have adopted 'Moleto,' while Revelli preferred 'Moleti.' Favaro, however, argued for 'Moletti,' on the grounds that this conforms to the usual way of forming Italian surnames and that it was the form in Rolls of the University of Padua ('Amici e corrispondenti di Galileo Galilei: XL. Giuseppe Moletti,' 47n). 'Moletti' was adopted by Rivolta, by Thomas B. Settle ('Galileo and Early Experimentation,' in *Springs of Scientific Creativity: Essays on Founders of Modern Science*, ed. Rutherford Aris, H. Ted Davis, and Roger H. Stuewer [Minneapolis: Univ. Minnesota Press, 1983], 3–20), and by me in previous articles, and I have kept it here.

44 Riccoboni, *Orationum volumen secundum*, ff. 43v–44r.

45 Giuseppi Moletti, *Discorso ... nel quale si dichiarano tutti i termini et le regole appartenenti alla Geografia*, printed with Ptolemy, *La Geografia*, tr. Girolamo Ruscelli (Venice, 1561); ed. Gio. Malombra (Venice, 1564); and printed separately (Venice, 1564 and 1573); *L'Efemeridi per anni XVIII ... 1563–1580* (Venice, 1563), reissued in 1564 for the years 1564–84 (Venice, 1564). On these works see Favaro, 'Amici e corrispondenti di Galileo Galilei: XL. Giuseppe Moletti,' 51, 54–5.

46 Venice, Archivio di Stato, Sezione Notarile: Testamenti, busta 646. Notaio

Michieli Francesco, Testamento 441; printed in Favaro, 'Amici e corrispondenti di Galileo Galilei: XL. Giuseppe Moletti,' 102–4.

47 The full title of what I call here the *Discourse on How to Study Mathematics* (or *Discorso sulla matematica*) is 'Discorso di M. Gioseppe Moleto Mathematico nel quale egli mostra che cosa sia matematica, quante sien le parti di quella, quali sieno, et come sono insieme ordinate, si discorre intorno a ciascuna, et insegna la via con la quale si debbano studiare per potersene impadronire; dichiara ancora in esso molti luoghi de filosofi et de mathematici, et insieme solve molte dubitationi, et scuopre molti secreti' (Milan, Biblioteca Ambrosiana MS. S 103 sup., f. 122r). Dedicated to a certain Giorgio Gozzi, it was written sometime after 1570: Moletti mentions the fire in the Venetian arsenal of 1569 (f. 166v) and a booklet by Mahometto on the division of figures newly printed by Commandino (f. 134v). 'Mahometto' was Machometus Bagdedinus, whose *De superficierum divisionibus* was printed by Federico Commandino in 1570 from a manuscript brought to him by John Dee, which is now Milan, Biblioteca Ambrosiana MS. P 236 sup. (Rose, *Italian Renaissance of Mathematics*, 199–200). Carugo dates the *Discors.* to between 1573 and 1575 on unspecified internal evidence ('L'Insegnamento della matematica,' 178).

48 Moletti's surviving letters to the Gonzaga court are in Mantua, Archivio di Stato, Archivio Gonzaga, serie E. XLV. 3, buste 1511, buste 1512, buste 1513, and buste 1514; and serie F. II. 8, buste 2612 and buste 2617; the letter of thanks is dated 14 January 1582 and is in serie E. XLV. 3, buste 1513.

49 Giuseppe Moletti, *Tabulae Gregorianae* and *De Corrigendo Calendario* [printed together] (Venice, 1580); see Favaro, 'Amici e corrispondenti di Galileo Galilei: XL. Giuseppe Moletti,' 68–84; Moletti's letters to Cardinal Sirleto concerning the reform of the calendar are preserved in Vatican City, Biblioteca Apostolica Vaticana MSS Vat. lat. 6194, f. 416, and Vat. lat. 6195, ff. 10, 12, 14, 102, 562, 680, and 684.

50 Carugo, 'L'Insegnamento,' 176–7; Favaro, 'Amici e corrispondenti di Galileo Galilei: XL. Giuseppe Moletti,' 63; see Milan, Biblioteca Ambrosiana MS. S 100 sup., ff. 1–90 (lecture notes on various works on optics); ff. 212–14 (Sacrobasco, *Sphere);* ff. 235–40 (Euclid, *Elements).*

51 *In librum Mechanicorum Aristolelis expositio,* Milan, Biblioteca Ambrosiana MS. S 100 sup., ff. 154–210 (hereafter cited as *Expositio);* the first sheet is dated, in Moletti's customary way, '1581 Oct. d. 6. hor. 14 1/2 in circa. Repetita 1582 10 febr.' (f. 156r); the University Rolls list Moletti as lecturing in 1585–6 on 'Lib. Euclidis et Mechan. Arist.' (Favaro, 'Amici e corrispondenti di Galileo: XL. Giuseppe Moletti,' 63; and Carugo, 'L'Insegnamento della matematica,' 176).

52 'tumultaria et ex tempore' (Moletti, *Expositio*, f. 156r); see also, e.g., 'audietis in solutione proprii Problematis' (f. 161r); and 'atque exponemus problemata

Philosophi' (f. 170v). The fourteen topics are listed on f. 156r–v, and on f. 197r there is a rough grouping of questions according to their subject matter. On the lectures see Rose, 'Professors of Mathematics at Padua,' 303, and *Italian Renaissance of Mathematics*, 287; and Carugo, 'L'Insegnamento della matematica,' 177.

53 Two very similar versions of the preface are found in Milan, Biblioteca Ambrosiana MS. S 100 sup., f. 196r (first version) and f. 194r (revised version); a third, somewat different version is in Milan, Biblioteca Ambrosiana MS. D 235 inf., f. 35r.

54 Riccoboni, *Orationum volumen secundum*, f. 43r; quoted in Favaro, 'Amici e corrispondenti di Galileo Galilei: XL. Giuseppe Moletti,' 95, and in Carugo, 'L'Insegnamento della matematica,' 178.

55 On Pinelli's library see Marcella Grendler, 'A Greek Collection in Padua: The Library of Gian Vincenzo Pinelli (1535–1601),' *Renaissance Quarterly* 33 (1980): 386–416, esp. 388–90, 402; and Rivolta, *Catalogo dei codici Pinelliani dell' Ambrosiana*.

56 Milan, Biblioteca Ambrosiana MS. S 100 sup., ff. 294r–318r. The *Dialogue* is listed among Moletti's works and briefly described in Carugo, 'L'Insegnamento della matematica,' 178; Carugo also notes the fragment in D 235 inf. mentioned below.

57 Milan, Biblioteca Ambrosiana MS. D 235 inf., ff. 59r–62v.

58 Giambatista Venturi, *Memorie e lettere inedita finora o disperse di Galileo Galilei* 1 (Modena, 1818); Raffaello Caverni, *Storia del Metodo Sperimentale in Italia*, 6 vols. (Florence, 1891–1900; rpt. New York: Johnson Reprint, 1972), 4:271–4, 290–2; see Favaro, 'Amici e corrispondenti di Galileo Galilei: XL. Giuseppe Moletti,' 88–90; and Carugo, 'L' Insegnamento della matematica,' 170.

59 Thomas B. Settle, 'Galileo and Early Experimentation,' in *Springs of Scientific Creativity: Essays on Founders of Modern Science*, ed. Rutherford Aris, H. Ted Davis, and Roger H. Stuewer (Minneapolis: Univ. Minnesota Press, 1983), 3–20, at 10–12.

60 Caverni, *Storia*, 271; Favaro, 'Amici e corrispondenti di Galileo Galilei: XL. Giuseppe Moletti,' 89.

61 On Franceschino see Pierre M. Tagmann and Michael Fink, 'Rovigo, Francesco (Franceschino),' *The New Grove Dictionary of Music and Musicians* (London: Macmillan, 1980), 16:279–80; and Iain Fenlon, *Music and Patronage in Sixteenth-Century Mantua* (Cambridge: Cambridge Univ. Press, 1980), 108 (my thanks to Jeffrey G. Kurtzman for these references).

62 One possibility is a colonel [Cesare?] Andreasi, who was consulted by Duke Guglielmo concerning the fortifications at Monferrato and Alba and who was apparently in Mantua in July of 1576, a few months before Moletti began

the *Dialogue;* on this Andreasi, see A. Bertolotti, *Architetti, ingegneri, e matematici in relazione coi Gonzaga, Signori di Mantova, nei secoli xv, xvi, e xvii* (Genova, 1889), 55–6 (my thanks to Thomas B. Settle for this suggestion and reference).

63 'A tutto questo della velocità, e tardità del movimento del diametro si puo giognere non picciolo discorso, il quale hora lasso, riserbandomi di farlo in altro luogo' (Moletti, *Discorso*, f. 162r).

64 For a discussion of Tartaglia's treatment, see W.R. Laird, 'Patronage of Mechanics and Theories of Impact in Sixteenth-Century Italy,' in Bruce T. Moran, ed., *Patronage and Institutions: Science, Technology and Medicine at the European Court, 1500–1750* (Woodbridge: Boydell, 1991), 51–66, at 58–60.

65 On the various meanings of 'machine' in antiquity and the sixteenth century, see Gianni Micheli, *Le origini del concetto di macchina*, Biblioteca di Physis 4 (Leo S. Olschki: Florence, 1995); on Archimedes' reputation as the defender of Syracuse, see W.R. Laird, 'Archimedes among the Humanists,' *Isis* 82 (1991): 629–38, esp. 636–7 for Moletti.

66 *Discorso*, ff. 169v–170v.

67 *Expositio*, ff. 163r–164v.

68 *Expositio*, ff. 165r–167v.

69 *Discorso*, ff. 157r–162r; the diagram mentioned is on f. 160r, and in Leonico, f. xxvii recto.

70 Pseudo-Aristotle, *Mechanical Problems*, 843b10–23; trs. Leonico, f. xxvi recto.

71 Nicholas Copernicus, *De revolutionibus orbium coelestium*, book 3, prop. 4, ed. Jerzy Dobrzychi, tr. Edward Rosen, *Nicholas Copernicus Complete Works*, 3 vols. (Warsaw/Cracow: Polish Academy of Sciences, 1972–85), 2:125–6; on al-Tûsî and the Tûsî-couple, see E.S. Kennedy, 'Late Medieval Planetary Theory,' *Isis* 57 (1966): 365–78, at 369–70; and see also Rosen's note to his translation, *De revolutionibus*, 384–5. The Prince here holds to the common view about Copernicus before Galileo, a view that had been advanced by Osiander in his (anonymous) Prologue to the *De revolutionibus;* on the early reception of Copernicus see Robert S. Westman, 'The Astronomer's Role in the Sixteenth Century: A Preliminary Study,' *History of Science* 18 (1980): 105–45. Moletti's papers contain numerous notes mentioning Copernicus: he began (in 1584) to write a commentary on the *De revolutionibus* (in Milan, Biblioteca Ambrosiana MS. D 235 inf., ff. 28r–29r) and he contemplated compiling a new ephemerides from Copernicus's tables, since they corresponded better to appearances than the Alphonsine Tables (see ff. 3r–9v, 82r, and f. 15r–v). Moletti's most extensive discussion of Copernicus occurs in his *Discourse on How to Study Mathematics*, where he recommends that Copernicus be studied even though his supposition that the earth moves and the sun stands still is

false, since he is able to save the appearances. Moletti sees the Copernican hypothesis simply as another mathematical device, equivalent to the alternatives (which he says are not 'entirely natural') proposed by Ptolemy, that either concentrics (i.e., deferents) and epicycles, or eccentrics alone, can equally well save the planetary appearances (*Discorso*, f. 147r); on Moletti and Copernicus, see Carugo, 'L'Insegnamento della matematica,' 174–5.

72 Pseudo-Aristotle, *Mechanical Problems*, 848b34–5; Leonico, ff. xxvii verso–xxviii recto; Piccolomini, ff. 12v–13v.

73 Moletti, *Discorso*, ff. 160v–162r.

74 Cf. 'It [i.e., the weight] is heavier in descending, to the degree that its movement toward the center (of the world) is more direct,' Jordanus, *De ratione ponderis*, ed. and tr. E.A. Moody and Marshall Clagett, in *The Medieval Science of Weights* (Madison: Univ. Wisconsin Press, 1952), 175; and 'Also we request that it be conceded that a heavy body in descending is so much the heavier as the motion it makes is straighter toward the center of the world,' Niccolò Tartaglia, *Quesiti*, book 8, qu. 24, 3rd petition; tr. Drake and Drabkin, *The Science of Mechanics in Sixteenth-Century Italy*, 118, where a diagram similar to Moletti's is found. Cf. Moletti's similar treatment in the *Discorso*, ff. 157r–v.

75 Tartaglia, *Quesiti*, book 8, qu. 24, 3rd petition; tr. Drake and Drabkin, *The Science of Mechanics in Sixteenth-Century Italy*, 119, where again a similar diagram is found.

76 See Moletti's similar discussion in the *Discorso*, f. 158r–v.

77 See the *Discorso*, ff. 158v–159r; for Euclid's proposition and the controversies surrounding the horn angle, see Sir Thomas L. Heath, *The Thirteen Books of Euclid's Elements*, 3 vols., 2nd ed. (Cambridge: Cambridge Univ. Press; rpt. New York: Dover, 1956) 2:39–43.

78 *Discorso*, f. 156r–v.

79 Cf. 'Cur virga longius mittatur a puero quam a viro investigare' (Girolamo Cardano, *Opus novum de proportionibus*, book 5, prop. 113 [Basel, 1570], and in *Opera* [London, 1663; rpt. in facsimile New York: Johnson Reprint, 1967], 4:517); in this proposition Cardano gives the reason, among others, that light things cannot receive as great an *impetus* as heavy things. Galileo would later make much the same point in an undated fragment on mechanics (*Opere di Galileo Galilei*, ed. Antonio Favaro, 23 vols. (Florence: Barbèra, 1891–1909), 8:572–3; tr. in I.E. Drabkin and Stillman Drake, *Galileo on Motion and Mechanics* [Madison: Univ. Wisconsin Press, 1960], 140).

80 See above, 10.

81 Aristotle, *De motu animalium*, esp. 698a7–699a12; tr. Leonico, xv verso–xvi recto.

82 Pseudo-Aristotle, *Mechanical Problems*, 858a23–b3, ed. and tr. Hett, 406–9.
83 For a more extensive discussion of Moletti's idea of the role of the resistance in impact, see Laird, 'Patronage of Mechanics and Theories of Impact,' 60–1.
84 Aristotle's view, expressed in *Physics*, 8.10, 266b27–267a12, that the air or other medium kept projectiles in motion, was much criticized by the ancient and medieval commentators. AN alludes to the most significant alternative to Aristotle's opinion developed in the Middle Ages – impetus theory. Impetus was conceived to be a motive virtue or power impressed as a form on the projectile by the mover. On impetus theory see Anneliese Maier, *Zwei Grundprobleme der scholastischen Naturphilosophie*, 2nd ed. (Rome: Storia e Letteratura, 1951); and 'Die naturphilosophie Bedeutung der scholastischen Impetustheorie,' *Scholastik* 30 (1955): 321–43; rpt. in Maier, *Ausgehendes Mittelalter* 1 (Rome: Storia e Letteratura, 1964), 353–79; tr. Steven D. Sargent, in *On the Threshold of Exact Science: Selected Writings of Anneliese Maier on Late Medieval Natural Philosophy* (Chicago: Univ. Chicago Press, 1982), 76–102; Marshall Clagett, *Science of Mechanics*, 505–40; and Ernest Moody, 'Galileo and His Precursors,' in *Studies in Medieval Philosophy, Science, and Logic* (Berkeley: Univ. California Press, 1976), 393–408.
85 For a translation and discussion of part of this passage see Settle, 'Galileo and Early Experimentation,' 10–12.
86 Cardano, *Opus novum de proportionibus*, book 5, prop. 110 (London, 1663; rpt. 1967), 4:515–16; Cardano's proposition is reprinted and translated in Lane Cooper, *Aristotle, Galileo, and the Tower of Pisa* (Ithaca: Cornell Univ. Press, 1935), 74–7.
87 For the history of this much-discussed question and for representative texts, see Pierre Duhem, *Études sur Léonard de Vinci*, 3 vols. (Paris: A. Hermann, 1906–13), 3:54–112; and Clagett, *Science of Mechanics*, 541–82. Note that Moletti does not include impetus among his alternatives, which was the favoured explanation from Buridan onwards.
88 John Buridan, *Questions on the Four Books on the Heavens and World of Aristotle*, tr. Clagett, *Science of Mechanics*, 558; Gasparo Cardinal Contarini (1483–1542) in his *De elementis* attributed to 'certain physicians' the opinion that all of nature is directed by an intelligence, so that heavy bodies can know to exert more effort the closer they come to their natural place – see Duhem, *Études sur Léonard de Vinci*, 3:183.
89 This seems a conflation of several views commonly expressed in the Middle Ages: see Duhem, *Études sur Léonard da Vinci*, 3:41–2, 432–4.
90 See Rose and Drake, 'The Pseudo-Aristotelian *Questions of Mechanics*'; and Laird, 'The Scope of Renaissance Mechanics.'

91 Tartaglia, *Quesiti et invenzione diverse*, 8, qu. 35; trans. Drake and Drabkin, *Mechanics in Sixteenth-Century Italy*, 32.

92 *Expositio*, ff. 186r–v.

93 *Expositio*, ff. 195r–v, 199r; this is apparently the second version of this section, the first (f. 169r–v) arguing to the same conclusion by distinguishing in the intermediate sciences a primary, intrinsic end (knowledge of causes and truth) from a secondary, extrinsic end (productive work).

94 *Expositio*, ff. 175r–178v.

95 *Expositio*, f. 180r.

96 Niccolò Tartaglia, *Nova Scientia* (Venice, 1537); *Quesiti*, books 1–4 concern ballistics, 5 and 6 fortification, 7 and 8 mechanics, and 9 mathematics; Francesco Maurolico, *Problemata mechanica cum appendice*, ed. Silvestro Maurolico (Messina, 1613), 10 (see Marshall Clagett, *Archimedes in the Middle Ages*, 3.3 [Memoirs of the American Philosophical Society, 125 B] [1978], 784–5); Galileo, 'Delle meccaniche lette in Padova l'anno 1594 da Galileo Galilei,' ed. Antonio Favaro (*Memorie del Reale Istituto Veneto di Scienze, Lettere ed Arti*, 26.5) (Venice: Carlo Ferrari, 1899); Stillman Drake, 'Galileo Gleanings, V: The Earliest Version of Galileo's *Mechanics*,' *Osiris* 13 (1958): 262–90, on 270; and Laird, 'Patronage of Mechanics and Theories of Impact.'

97 Laird, 'Scope of Renaissance Mechanics,' 56–8.

98 Baldi, *Life of Archimedes;* this passage is translated in Drake and Drabkin, *Mechanics in Sixteenth-Century Italy*, 14–15.

99 Bernardino Baldi, *Mechanica Aristotelis problemata exercitationes* (Mainz, 1621), ff.):():(r–):():(2v; see Rose, *Italian Renaissance of Mathematics*, 248–51.

100 Guidobaldo del Monte, *Mechanicorum liber* (Pesaro, 1577); Guidobaldo, *Le mechaniche*, tr. Filippo Pigafetta (Venice, 1581); both rpt. Venice, 1615; German edition, *Mechanischer Kunst Kammer erster Theil*, tr. Daniel Mögling (Frankfurt, 1629). An abridged translation appears in Drake and Drabkin, *Mechanics in Sixteenth-Century Italy*, 239–327. On Guidobaldo's life and works see Paul Lawrence Rose, 'Monte, Guidobaldo, Marchese del,' *DSB*, 9:487–9; Domenico Bertoloni Meli, 'Guidobaldo dal Monte and the Archimedean Revival,' *Nuncius* 7.1 (1992): 3–34; and on the *Mechanicorum liber* see Wallace, *Galileo and His Sources*, 206.

101 Pappus's *Mathematical Collection*, book 8 of which was devoted to a treatment of simple machines largely derived from Hero, had been translated into Latin by Federico Commandino sometime before 1575, and although not printed until 1588, this translation was available in manuscript to Guidobaldo, formerly a student of Commandino's. On Pappus in general and Commandino's translation in particular, see Marjorie Nice Boyer,

'Pappus Alexandrinus,' in *Catalogus translationum*, ed. Kristeller and Cranz, 2:205–13, esp. 207–8; and 3:426–31, esp. 426–8.

102 Rose, *Italian Renaissance of Mathematics*, 222–42; and Drake and Drabkin, *Mechanics in Sixteenth-Century Italy*, 44–8.

103 A copy of the theorems, followed by attestations of their originality and Moletti's appraisal, is in Milan, Biblioteca Ambrosiana MS. A 71 inf., ff. 95r–96r (where Pinelli has attributed them to Vincenzo Galilei, Galileo's father); Moletti's appraisal reads as follows: 'A di 29 di Decembre del 1587. Io Gioseppe Moleto lettor publico delle Mathematiche nello studio di Padova, dico haver letti i presenti lemma, et theorema, i quali mi son parsi buoni, estimo l' autor d' essi, esser buono, et esercitato geometra. Il medesimo Gioseppe ha scritto di man proprio'; the theorems and the other appraisals are printed in Galileo, *Opere* 1:183.

104 Milan, Biblioteca Ambrosiana MS. D 235 inf., f. 25r (propositions 2 and 3) and f. 24r–v (proposition 13).

105 *Opere* 19:120; Rose and Drake, 'The Pseudo-Aristotelian *Questions of Mechanics*,' 94.

106 Stillman Drake, 'Galileo Gleanings, V: The Earliest Version of Galileo's *Mechanics*,' *Osiris* 13 (1958): 262–90, esp. 270; see also Antonio Favaro, 'Delle meccaniche lette in Padova l'anno 1594 da Galileo Galilei,' *Memorie del Reale Istituto Veneto di scienza, lettere ed arti*, 26.5 (Venice: Carlo Ferrari, 1899); for a fuller discussion of Galileo's early ideas of mechanics, see Laird, 'Scope of Mechanics,' 62–3.

107 Galileo Galilei, *Le meccaniche*, in *Opere* 2:155–90; tr. Stillman Drake, in Drabkin and Drake, *Galileo Galilei on Motion and Mechanics*, 147–82; see also Drake's introduction, 137–43.

108 Galileo Galilei, *De motu (On Motion)*, in *Opere* 1:251–366; tr. Drabkin, in Drabkin and Drake, *Galileo Galilei on Motion and Mechanics*, 13–123; *De motu dialogus (Dialogue on Motion)*, *Opere*, 1:368–408; tr. Drabkin, in Drake and Drabkin, *Mechanics in Sixteenth-Century Italy*, 329–77; the memoranda are found in *Opere* 1:409–17 and are translated in Drake and Drabkin, *Mechanics in Sixteenth-Century Italy*, 378–87. On the *De motu antiquiora* see I.E. Drabkin, 'A Note on Galileo's *De motu*,' *Isis* 51 (1960): 271–7; Raymond Fredette, 'Galileo's "De motu antiquiora,"' *Physis* 14 (1972): 321–50; and Michele Camerota, *Gli Scritti 'De motu antiquiora' di Galileo Galilei: Il Ms Gal 71* (Cagliari: CUEC Editrice, 1992).

109 See Stillman Drake, *Galileo at Work: His Scientific Biography* (Chicago and London: Univ. Chicago Press, 1978), 102.

110 For descriptions of the Ambrosiana manuscripts, see Astrik L. Gabriel, *A Summary Catalogue of Microfilms of One Thousand Scientific Manuscripts in the*

Ambrosiana Library, Milan (Notre Dame, Ind.: Mediaeval Institute, 1968), #173 and #894; Paolo Revelli, *I codici Ambrosiani di contenuto geografica* (Milan, 1929), #65 and #372; A. Rivolta, *Catalogo dei codici Pinelliani (latini) dell' Ambrosiana* (Milan, 1933), 192 and 250.

111 See 20–1 above.

Di Giuseppe Moleto. — Adi p.o d'Ottobre del 76. Alt.e ☉ gr. 17. hor 14. m. 33.
Cose tutte imperfette et senza risolutione certa.

S. AN. Queste tante machine et questi tanti stromenti ch'io veggo in questo luogo di V. A. mi fanno uenire a mente una
disputa ch'io udi fare un giorno ad alcuni Ingegneri, et Architetti insieme intorno a molte forze. Si domandauano
l'un l'altro diuerse cose, et tra le principali l'uno di loro ricercaua all'altro la cagione perche due pezzi d'Artigliaria ine-
guali ma così conditionati che l'uno d'essi porti minor balla
ma sia piu lungo dell'altro tirerà di mira piu lontano et farà
meno botta dell'altro, et l'altro rispondeua non so che di centro
et di circonferenza, et diceua che'l piu lungo descriue maggior
cerchio, et ueniua a mio giuditio intricando la cosa in modo
che meno s'intendeua dell'ultimo di quello, che prima si face-
ua. l'Altro poi replicaua, chiedendo la cagione perche con le taglie si leua così gran peso: et ricercaua insieme che le moltiplicasse la forza delle girelle delle taglie, et diceua io hò due taglie
con due girelle per taglia, et la maggior forza che fanno tirar
da uno ò piu huomini è di leuar tanto peso. giungo all'una
et all'altra una girella ò a tutte due domando quanta forza
s'hauerà accresciuto tirati però dallo stesso dell'istessi huomini, sopra che si dissero molte cose et in somma ò fosse perch'io
non intendessi bene le loro ragioni, ò come si uoglia che fosse io
ne restai poco sodisfatto. Si passò poi d'un ragionamento nell'altro
fin che si uenne alla cosa delle fortezze, et delle mine doue
si discorsero molte cose di non picciolo momento. Son certo che
se l'A. V. l'hauesse udite ne hauerebbe hauuto contento. PR.
hauerei hauuto contento per certo, et tanto maggiore quanto
se gli huomini che di tali cose discorreuano intendeuano l'arte

Plate 1: The *Dialogue on Mechanics*, Milan, Biblioteca Ambrosiana MS. S 100 sup., f. 295r. Reproduced with permission. Property of the Ambrosiana Library.

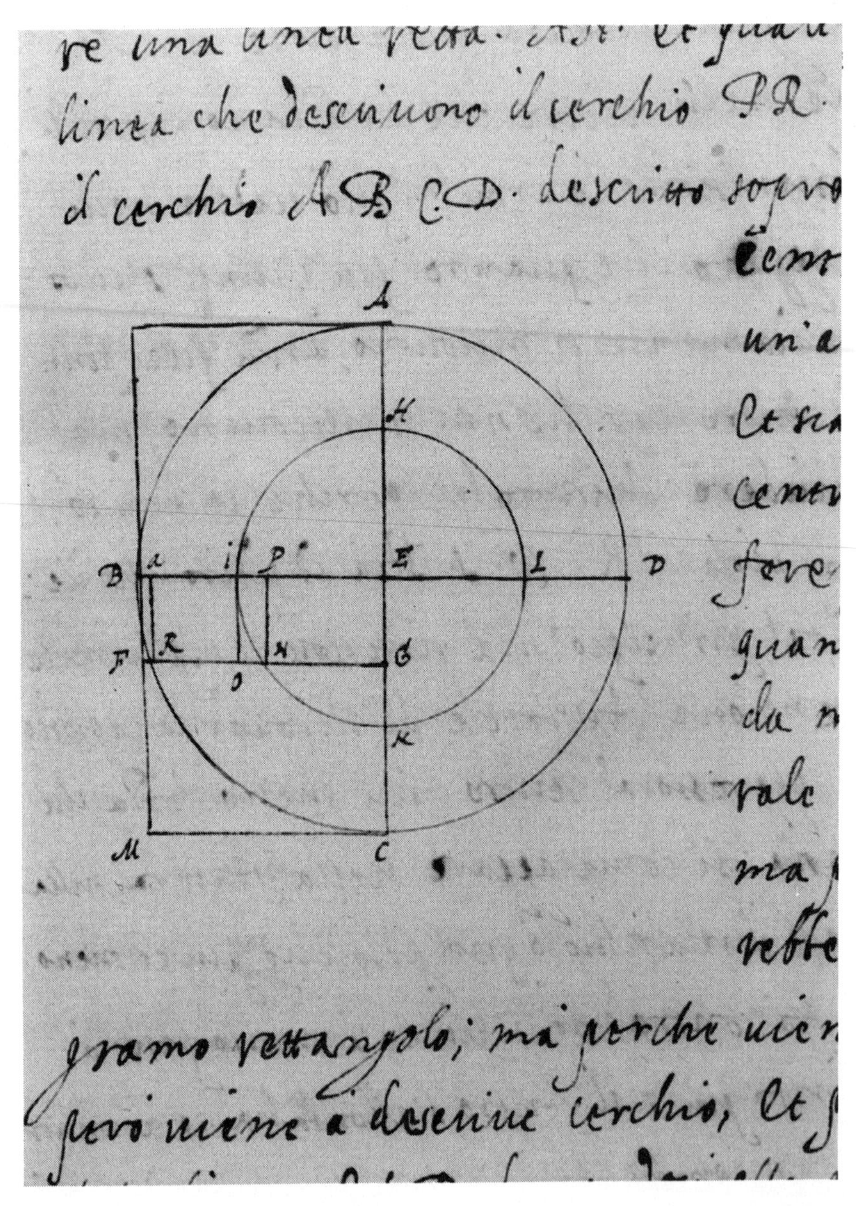

Plate 2: The principle of circular movement, Milan, Biblioteca Ambrosiana MS. S 100 sup., f. 301v. Reproduced with permission. Property of the Ambrosiana Library.

del matematico, seguirà che la mecanica sia subalternata alla 59
matematica. Oltre quello che si è detto si aggiunge il testimonio
di Proclo autore gravissimo, il quale la mette tra le parti
della matematica. Come V.S. sa, poi le scientie nella loro
prima divisione sono di due sorti, percioche o sono
prattiche, o contemplative: et lasciando le prattiche ca-
pare diremo delle contemplative, le quali sono tre, ~~[illegible]~~
~~[illegible]~~ cioè la metafisica, la matematica, et la
fisica o naturale. ne possono essere altre scientie oltre
queste, essendo che se altre ne ne sono, sono dalle dette
dependenti. La sufficienza della prima divisione potriano
noi venir deducendo così: L'huomo composto di corpo ~~[illegible]~~
et di anima, et come veggiamo, et quanto all'una parte, et quanto
all'altra è imperfetto, et però ha bisogno trovar vie che possino
ridurr'a perfettione l'una et l'altra parte, ~~[illegible]~~ il che da' savij
si fa con l'aiuto delle scientie, così prattiche come contemplative. Il fine adunque delle scientie prattiche è stato di ridurre
a perfettione l'operationi dell'huomo. et il fine delle ~~[illegible]~~
contemplative di ridurre a perfettione la parte intellettiva di
quello. Et però la diffinitione di queste due scientie è che la
prattica ha per fine l'operatione, et la contemplativa la cognitione. Ma non ogni operatione è fine del prattico, percioche quando
così fusse, l'operationi inique et laide sarebbono fine del prattico
ma solo l'operationi virtuose, et buone, hanno da essere fine
della scientia prattica: et il fine essendo qual vuole il bene, et il vero
et non l'apparente, però il bene possibile è fine del prattico. Così
ancora non ogni ~~[illegible]~~ cognitione ha da essere ~~[illegible]~~ il fine de-
contemplativo, ma solo la ~~[illegible]~~ cognitione della verità
et però la scientia contemplativa ~~[illegible]~~ ha per fine
il vero

Plate 3: The fragment on mathematics in Moletti's hand, Milan, Biblioteca Ambrosiana MS. D 235 inf., f. 59r. Reproduced with permission. Property of the Ambrosiana Library.

THE *DIALOGUE ON MECHANICS*

Sigla

D = Milan, Biblioteca Ambrosiana MS. D 235 inf., ff. 59r–62v (1576).

S = Milan, Biblioteca Ambrosiana MS. S 100 sup., ff. 294r–318r (1576).

G = Florence, Biblioteca Nazionale Centrale, Galileo MS. 329, ff. 3r–9v (*ante* 1818).

canc. cancellavit

corr. correxit

del. delevit

ins. inseruit

om. omisit

scrip. scripsit

< > omissionem vel capitulum supplevimus

<Dialogo Intorno alla Meccanica>[1]

di Giuseppe Moleto[2]

<Giornata Prima>

<IL SOGGETTO ED I PRINCIPII DELLA MECCANICA>

Signor AN. Queste tante machine et questi tanti stromenti ch' io veggo in questo luogo di Vostra Altezza mi fanno venire à mente una disputa ch' io udi fare un giorno ad alcuni ingegneri et architetti insieme intorno à molte forze. Si domandavano l' un l' altro diverse cose, et tra le principali, l' uno di loro ricercava all' altro la cagione perché due pezi d' artiglieria inequali ma così conditionate che l' uno d' essi porti minor balla, ma sia più lungo dell' altro, tirerà di mira più lontano, et farà meno botta dell' altro. Et l' altro rispondeva, non so che di centro et di circonferenza, et diceva che 'l più lungo descrive maggior cerchio, et veniva à mio giuditio intricando la cosa in modo che meno s' intendeva nell' ultimo di quello che prima si faceva. L' altro poi replicava chiedendo la cagione perche con le taglie si leva così gran peso: et ricercava insieme che le moltiplicasse la forza della girelle delle taglie, et diceva: «Io ho due taglie con due girelle per taglia, et la maggior forza che fanno tirati da uno ò più huomini è di levar tanto peso. Giungo all' una et all' altra una girella ò à tutte due: domando quanta forza l' haverò accresciuto, tirati però dallo stesso dell' istessi huomini». Sopra che si dissero molte cose et in soma, ó fosse perch' io non intendessi bene le loro ragioni ó come si voglia che fosse, io ne restai poco sodisfatto. Si passo poi d' un ragionamento nell' altro, fin che si venne alla cosa delle forteze et delle mine, dove si discorsero molte cose di non picciolo momento. Son certo che se l' Altezza Vostra l' havesse udite ne haverebbe havuto contento.

1 di Giuseppe Moleto alcune memorie in materia d'artiglieria *manu Pinelli S, 294r*

2 Di Giuseppe Moleto *manu Pinelli S, 295r* / À di primo d' ottobre del 76, altitudo solis gradus 17 hora 14 minuta 33 *S* / Cose tutte imperfette et senza risolutione certa. *S*

<Dialogue on Mechanics>

by Giuseppe Moletti

<The First Day>

<THE SUBJECT AND PRINCIPLES OF MECHANICS>

Signor AN. These many machines and these many instruments I see here belonging to Your Highness bring to mind a dispute that I heard one day among some engineers and architects concerning many forces. They were asking one another various things, and one of the main things was that one of them asked another the reason why, of two pieces of artillery, the one that takes a smaller ball but is longer than the other will shoot farther and will make less impact. And the other replied something about centre and circumference, and said that the longer shot describes a greater circle, and in my opinion complicated the thing in a way that it meant less at the end than at the beginning.[1] The first then asked the reason why one can lift such a great weight with a pulley; and he also asked what multiplies the force of the wheels of the pulley, and he said: 'I have two pulleys, each with two wheels, and the greater force they produce pulled by one or more men can lift so much weight. I add one wheel to the one or to the other, or to both: I then ask, how much shall I have increased the force, while still pulled by the same <number> of the same men?'[2] They said many things about this and in the end, either because I did not follow their arguments very well or whatever, I was still little satisfied. The discussion then passed from one argument to another, and afterwards it turned to matters concerning fortification and mines, where many things were discussed of no little moment.[3] I am certain that if you had heard it you would have been satisfied.

PR. Haverei havuto contento per certo, et tanto maggiore quanto se gli huomini che di tali cose discorrevano intendevano l' arte [S, 295v] et i principii delle cose, che tra loro trattavano. Le questioni che Vostra Signoria Illustrissima ha raccontate sono di molto momento et porgono à professori delle scienze non picciola difficoltà. Et male si possono sciogliere senza l' havere inteso bene i principii dell' arte che di queste cose tratta.

<I L' arte del ingegnero e la scienza della meccanica>

S. AN. Et quale è l' arte che tratta et considera le questioni dette di sopra?

PR. È quell' arte che si dice mechanica, et noi potremo dirli l' arte dell' ingegniero.

AN. S' ell' è l' arte dell' ingegniero, et quei detti di sopra erano ingegnieri, adunque potevano benissimo solvere le sudette questioni.

PR. Vostra Signoria Illustrissima conchiude, che potevano, essendo ingegnieri, solvere le sudette questioni: ma secondo il parere di Vostra Signoria non l' hanno sciolte; adunque può dire che non fossero ingegnieri.

S. AN. Come no, sono pure professori di tal arte et per tali conosciuti; adunque sono ingegnieri.

PR. Et quanti sono che fanno professione de medici et sono conosciuti per tali, et nondimeno niuna cosa intendono meno che la medicina? Et quel ch' io dico della medicina posso dir di tutte le professioni. Si che non vane[3] sono chiamati ingegnieri, et conosciuti per tali adunque intendono l' arte dell' ingegnero. Anzi son di parere per Dio à Vostra Signoria il tutto che chi domandasse loro qual sia l' arte di che fanno professione non saprebon dirlo. Non dico non esservene alcuno che l' intenda, perché sarebbe cosa fuor d' ogni convenevolezza, che tra tanti ingegneri non ne fossero alcuni ch' intendessero l' arte loro per i veri termini, ma credo che sieno molto pochi.

S. AN. Quest' arte dell' ingegniero, è cosa che si possa imparare su i libri si come avviene di molte scienze, ò pure è una semplice prattica la quale s' impara col pratticare con gli huomini[4] che di quella fanno professione, si come è l' arte del sarto, del calzolaio, del marangone et tali? perché la maggior parte degl' ingegneri c' ho conosciuti sono quasi

3 vane] vale *S*
4 huomini] huomeni *S*

PR. I would certainly have been satisfied, and so much the more if the men who were discussing such things understood the art [S, 295v] and the principles of the subject that they were treating. The questions that Your Most Illustrious Lordship related are of great moment and give those who profess the sciences not a little trouble. And they can hardly be resolved without having well agreed on the principles of the art that treats of these things.

<I The Art of the Engineer and the Science of Mechanics>

AN. And what is the art that treats and considers such questions?

PR. It is the art called mechanics, and we could call it the art of the engineer.

AN. If it is the art of the engineer, and those mentioned before were engineers, then they could very well resolve those questions discussed.

PR. You conclude that they, being engineers, could resolve the aforementioned questions: but it appeared to you that they did not resolve them; so one could say that they were not engineers.

AN. But no, they are indeed masters of this art and are recognized as such; therefore they are engineers.

PR. And how many are there who make a profession as physicians and are known as such, and nevertheless understand nothing less than they do medicine? And what I say of medicine I could say of all the professions. If they are not called and known as engineers in vain, then they must understand the art of the engineer. In fact, I am emphatically of the opinion that everyone whom one might ask what the art is of which he makes a profession would not know how to describe it. I do not say there are not some who do understand it, because it would be beyond all expectation that among so many engineers there are not some who understand their art in its true terms, but I believe they are very few.

AN. This art of the engineer, is it something one can learn from books as is the case with many sciences, or is it just a simple skill that one learns with practice among the men who make a profession of it, as are the arts of the tailor, the shoemaker, the carpenter, and such? – because the greater part of the engineers that I have known are almost

senza lettere. Anzi ho cono[S, 296r]sciuto un Thomaso Scala stimato ingegnero di alcun momento che appena sapeva leggere. Et n' ho pratticati molti altri che non sapevano ne leggere ne scrivere.

PR. L' arte dell' ingegniero s' impara parte su i libri et parte pratticando, ma il principale fondamento suo si cava da libri.

S. AN. Et chi ha scritto di quella?

PR. Molti greci, come Aristotele, Atheneo, Pappo, Herone, Archimede, et altri molti. Tra latini, da Giordano un po'[5], et d' alcune cose che sono in Vitruvio, non saprei de gli antichi chi n' havesse scritto.

S. AN. Et che cosa è quest' arte delle ingegniero? consiste in altro che intorno alla cosa delle forteze, et del misurare i terreni, et condur acque? – poi che tutti gl' ingegnieri c' ho conosciuti facevano professione delle sopradette cose.

PR. Se Vostra Signoria sapesse à quanto si stende l' arte dell' ingegniero si stupirebbe.

A. Supplico Vostra Altezza à dirmene alcuna cosa accioche possa vedere l' inganno degl' ingegnieri d' hoggi.

P. Col dire à Vostra Signoria Illustrissima la definitione della mechanica messa d' Aristotele verrà da se à comprendere la eccellenza dell' arte dell' ingegniero, poiche <è> l' istessa che la mecanica. Dice Aristotele che così gli effetti naturali come quelli che contra l' ordine della natura dall' arte et dall' industria all' uso humano vengon prodotti, de quali non se ne sa la cagione, sono giudicati miracoli ò ci porgono grandissima maraviglia. Veggiamo noi la natura operare sempre ad un modo, il che essendo molte cose vengono prodotte da quella che sono molto lontane dall' utilità nostra, et contrariano al nostro uso. La dove dovendo l' huomo far alcuna operatione contra l' ordine naturale, come à far che le cose gravi vadino all' insù, tal cosa con la difficoltà sua ci porge non picciola ansieta, et fa che machinando troviamo modo di superare tali difficultà. Et però quell' arte che ci insegna à superare queste difficultà noi denominamo[6] dal machinamento mecanica. Et è certa cosa che noi con l' arte superiamo quello con che supera noi la natura.

[S, 296v] Or veda Vostra Signoria à quanti modi possiamo operare per utile nostro contra l' ordine naturale à tanti diremo destendendosi la mecanica. Ma accioche Vostra Signoria non habbia in simil cosa da dubitare, soggiongerò che le difficoltà, le quali debbono essere superate dal

5 un po'] in poi *S*
6 denominamo] domandiamo *S*

illiterate.[4] In fact, I knew [296r] a Thomaso Scala, considered an engineer of some repute, who scarcely knew how to read.[5] And I have dealt with many others who could neither read nor write.

PR. One learns the art of the engineer partly from books and partly by practice, but its principal foundations one gets from books.

AN. And who has written on it?

PR. Many Greeks, such as Aristotle, Atheneus, Pappus, Hero, Archimedes, and many others. Among the Latins, a little from Jordanus, and <apart> from some things that are in Vitruvius, I would not know of the ancients who have written on it.[6]

AN. And what is this art of the engineer? does it consist in anything but what concerns fortification, land surveying, and waterworks? – since all the engineers I have known made a profession of such things.

PR. If you knew how far the art of the engineer extended you would be amazed.

AN. I implore you to tell me something of it so that I can see the deceit of today's engineers.

PR. To give you the definition of mechanics as proposed by Aristotle would be to comprehend the excellence of the art of the engineer, since it is the same as mechanics. Aristotle says that natural effects, as well as those that oppose the order of nature and are produced by art or industry for human use, when we do not know their causes, are judged to be miracles or give rise to the greatest amazement. We see that nature always acts in one way, so that many things are produced by it that are very far from our utility and contrary to our use. For this reason, whenever man must do any work against the natural order, such as make heavy things go up, the difficulty of such a task brings no little anxiety, and compels us to find a way by machination to overcome such difficulties. And thus the art that teaches us to overcome these difficulties we call 'mechanics,' from 'machination.' And it is certain that 'we overcome with art that in which nature conquers us.'[7]

[S, 296v] Now you see that we can work in as many ways for our utility against the natural order as we can say by extension 'mechanics.' But in order that you not have any doubts, I shall suggest that the difficulties that are to be overcome by the mechanic must be overcome neither with an equal nor with a greater force. For in such cases the engineer would

mecanico, non hanno da superarsi ne con forze equali ne con maggiori. Poiché in simili casi l' ingegniero non meriterebbe lode alcuna: come se un' huomo movesse da terra un peso di 20 libre, ò come se tirasse all' insù con una corda un peso di 10 libre, ò come se un luogo fosse da 1000 fanti custodito contra altri 1000 ò meno. Ma s' haverà da laudare quando con poche forze supererà grandissime difficoltà: come che un' huomo solo, aiutato dall' ingegno del mecanico, muova un peso di[7] 1000 ò più libre; et come che l' istesso tiri all' insù con aiuto d' alcuno stromento altretanto peso con facilità così; ancora che 10 huomini difendano con l' aiuto delle machine un luogo contra 1000 ò 2000. Et in queste versa tutta ò la maggior parte del machinamento mecanico. Et da ciò vene infinità lode ad Archimedo, poich' egli solo diffese Siracusa per molto tempo contra il potentissimo esercito de romani, del quale tutti gli scrittori ne ragionavo con infinito honore suo et con meraviglia grandissima. Et del quale ragionando Plinio, scrittore di molta et grave autorità, dice: «Gran testimonio è stato della geometrica et machinale scienza di Archimede quello di Marco Marcello, il quale fece far grida che nel pigliarsi Siracusa solo egli fosse salvo et sarebbe successo se fosse stato da soldati conosciuto». Percioche mentre egli stava contemplando intorno ad alcune figure geometriche, vogliono che dicesse il soldato che l' era sopravenuto che non vi guastar queste figure. La dove stimando il soldato d' esser burlato, l' uccise, con molto dispiacere di Marcello, solito à dire che l' esercito romano haveva ricevuta più danno da Archimede solo che da tutti i soldati di Siracusa. Epilogando adonque à Vostra Signoria quello che [S, 297r] questa mecanica sia, dirò essere una scienza la quale insegna à superare[8] le difficoltà che contravengono all' uso humano con poche forze.

A. À me pare questa mecanica che Vostra Altezza ha nominato distendere i termini suoi tutte le arti che hoggi si usano tra noi, perché quale è quell' arte fattiva che non cerchi con poche forze di far molta operatione? – et per ciò mi par di dire ch' ella sia la regina delle arti mechaniche.

P. Vostra Signoria avverta una cosa che quantunque l' arte detta da noi sia chiamata mecanica, non dimeno è molto differente da quelle antiche mecaniche vengon dette. Anzi queste falsamente sono chiamate mecaniche, poi che più tosto «sellularie» ò basse et humile si debbon

7 *ante* di *del.* S i
8 superare] sapere S

not merit any praise: as if a man were to lift from the ground a weight of twenty pounds, or draw up with a rope a weight of ten pounds, or if a place were held by a thousand footsoldiers against another thousand or fewer. But it is praiseworthy when with small forces one overcomes the greatest difficulties: as when one man alone, helped by the ingenuity of the mechanic, moves a weight of a thousand or more pounds; and when the same man draws up with the help of an instrument as great a weight with ease; again, when ten men defend a place with the help of machines against one or two thousand.[8] And in these cases all or the greater part is due to the mechanical machinations.[9] And from this comes the infinite praise for Archimedes, because he alone defended Syracuse for a long time against the most powerful army of the Romans, which all the writers discuss with infinite honour and with the greatest amazement. And in discussing this, Pliny, a writer of great and solemn authority, says: 'Great testimony it was, to the geometry and mechanical science of Archimedes, that Marcus Marcellus ordered that in the capture of Syracuse he alone should be spared, and he would have been had he been recognized by the soldiers.'[10] For while he stood contemplating some geometrical figures, he reputedly told the soldier who approached him not to spoil these figures. Thinking that he was being mocked, the soldier killed him, to the great displeasure of Marcellus, who later often said that the Roman army had received more damage from Archimedes alone than from all the soldiers of Syracuse.[11] Adding then for you what [S, 297r] mechanics is, I say that it is a science that teaches one to overcome with small forces the difficulties that oppose human uses.[12]

AN. It seems to me that this mechanics that you have defined includes within its bounds all the arts that are used by us today, because what factive art does *not* seek with small forces to do great work? – and thus it seems to me that you have defined the queen of the mechanical arts.

PR. You allude to the fact that although the art we are discussing is called mechanics, nevertheless it is very different from what were called mechanical in antiquity. In fact, those were falsely called mechanical, because they should rather be called 'sellularian,' or base and low, than mechanical, and Aristotle called them 'banausiai,' which means <handi-

dire che mecaniche, et Aristotele le chiama «banausiai[9]», che vuol dire <artigianati[10]>. Bene però ha detto Vostra Signoria che la mecanica à tutte l' altre arti presta grandissimo giovamento, poiché à lei sola s' appartiene rendere la ragione di tutte quelle cose che con picciola forza fanno con l' aiuto suo che non potrebbe farsi senza lei con molta. Et quantunque il mecanico dia aiuto à tutte le arti, nondimeno egli non opera col corpo cosa alcuna, ma tutto machina con la mente et usa gli artefici come suoi stromenti, si come veggiamo fare à gli architetti, i quali solo ordinano à fornaciari, muratori, à gli scarpelini, à gli scultori, à marangoni, et à gl' altri artefici quello ch' hanno ad operare intorno ad alcuna fabrica. Et così apunto fanno coloro che ordinano hoggi le forteze, percioche basta loro dare i dissegni et le misure et vedere le qualità de siti sopra quali constituiscono non solo la pianta, ma l' impiego[11] della forza con tutte le misure necessarie et che alla perfettione si richieggono della cosa.

A. L' Altezza Vostra ha chiamata la mecanica hora col nome di arte, et hora col nome di scienza. Desidererei sapere che cosa ella fosse: perchioche se fosse arte non saprei vedere come non havesse il mecanico da operare, essendo l' arte un habito fattivo con ragione si come vuole Aristotele; che fosse scienza, non saprei ne ancora capire [S, 297v] come fosse cagione de operationi, essendo che le scienze consistono solo in contemplationi.

P. Vostra Signoria ha da sapere che così il nome di arte alcune volte si piglia per la scienza, come il nome della scienza molte volte si piglia per l' arte, et si trovano spesso appresso ad Aristotele stesso presi i sudetti nomi indifferentemente. Alla mecanica dunque si può dire l' uno et l' altro nome, preso però et l' uno et l' altro, come si dice largamente. Con tutto ciò restringendo la cosa, io dirò la mecanica esser più tosto scienza che arte, essendo che è sotto una scienza ò subalternata ad una scienza, et appresso ha quelle cose che si richiegono ad una scienza. Anzi non solamente è sotto una scienza ma sotto due, ciò è alla naturale et alla matematica. Che sia scienza si dimostra così: ogni habito di conclusione acquistato col mezo della dimonstratione è scienza, poi che tale è la definitione della scienza; ma la mecanica è un tal habito; adonque la mecanica è scienza. Che la mecanica sia un tale habito io ne so provandolo a Vostra Signoria di mano in mano. Che poi sia subalternata alla scienza

9 banausiai] vanausie *S; cf.* *βαναυσίαι*
10 artigianati *lacuna in S*
11 impiego] impie *S*

crafts>.[13] But you were right to say that mechanics lends all the other arts the greatest help, because it alone provides the reason for all those things that are done with its help with a small force that could not be done with a large force without it. And although the mechanic helps all the arts, nonetheless he does not do anything bodily, but devises everything in his mind and uses craftsmen as his instruments, just as we see architects do, who alone order blacksmiths, bricklayers, engravers, sculptors, carpenters, and the other craftsmen whose job it is to work on some building.[14] And this is exactly what those who design fortresses do now, because it is their job to make the designs and the measurements and to view the characteristics of the sites, where they put together not only the plan but also the deployment of forces with all the necessary measurements and what is required for the completion of the work.[15]

AN. Sometimes you have called mechanics an art, sometimes a science.[16] I should like to know which it is: because if it were an art, I do not see how the mechanic would not have to do work, since art is a factive habit with reason, as Aristotle says; if it were a science, I still do not understand [S, 297v] how it can be the cause of works, since sciences consist only in contemplation.[17]

PR. You need to know that just as the name art is sometimes used for science, so the name science is used many times for art, and one often finds in Aristotle himself the terms used side by side indifferently. So, speaking generally, one can call mechanics by either name or by both. More strictly, though, I would say that mechanics is rather science than art, since it is under or subalternated to a science, and consequently it has what pertains to a science. In fact, not only is it under a science but under two, viz. natural philosophy and mathematics.[18] That it is a science is demonstrated as follows: every habit of conclusion acquired by means of demonstration is science, since such is the definition of science; but mechanics is such a habit; therefore mechanics is a science. That mechanics is such a habit I know and I shall prove it to you in a moment. That it is subalternated to mathematics I can demonstrate with the authority of Aristotle, who says at the beginning of the *Mechanical Problems,* concerning these things of which we speak, 'they are not entirely natural or entirely different from the natural, but are common to the natural just as to the mathematical.'[19] For as to their demonstrations they

matematica, lo dimostro con l' autorità d' Aristotele, il quale dice nel principio delle *Mecaniche* cose, oltre à ciò queste cose delle quali parliamo, «non sono al tutto naturali ne al tutto da quelle diverse: ma sono communi così alle cose naturali come alle matematiche». Percioche quanto alle dimostrationi sono matematiche, et quanto al soggetto, intorno à che versano, sono naturali. Allo stesso modo viene ad essere l' astronomia et la perspettiva. Con tutto ciò, molti sono stati di parere, et con ragione, che la mecanica fosse più tosto matematica che naturale, essendo che se le scienze hanno da ricevere la denominatione loro dalla ragione formale,[12] la quale essendo in questo caso la demostratione, et le demostrationi delle mecaniche essendo quelle [D, f. 59] del matematico, seguirà che la mecanica sia subalternata alla matematica. À tutto quello che s' è detto si giugne il testimonio di Proclo autore gravissimo, il quale la mette tra le [S, 298r] parti della matematica. Appresso, la mecanica ha quelle cose, che si richiegono ad una scienza, adonque diremo essere scienza. Ha primieramente il suo soggetto le passioni del quale non si dimostrano da altra scienza, le quali il mecanico le dimostra di quello per mezo de suoi proprii principii, et però sarà scienza.

<II Il soggetto della meccanica>

AN. Et quale è il soggetto della mecanica?

PR. Il soggetto nelle scienze ha da essere quello del quale se ne vengono considerando le proprie passioni, si come il soggetto dell' astronomo è il movimento del cielo numerabile, poi che di quello viene à considerarne le proprietà. Così dirò che 'l soggetto del mecanico sarà la machina, da dove viene denominata la mecanica.

AN. Et come proverà l' Altezza Vostra essere la machina?

PR. Proverolo così. Si ricorda haver letto Vostra Signoria appresso à Plutarco, et à Livio l' assedio di Siracusa, et oltre à ciò quello che Archimede fece fare ad Hierone Re di Siracusa? Et quello che egli fece d' una sua nave secondo Plutarco?

AN. Mi ricordo, ma non bene.

PR. Narra Proclo che Hierone haveva fatto fare una gravissima nave per mandare à Tolemeo Re d' Egitto. Et fatta, che fu volendola mandare in aqua; non fu sufficiente tutto il populo di Siracusa a farlo. All' hora Archimede disse al Re di voler fare che egli solo la mandasse in acqua. Il Re si rise di questo. Archimede fece la machina et l' applico alla nave, et

12 formale] formate *S*

are mathematical, and as to the subject that they consider they are natural. The same applies to astronomy and optics. With all such it has seemed to many – and with reason – that mechanics is more mathematical than natural,[20] since if sciences receive their denomination from their formal principle, which in this case is demonstration, and the demonstrations of mechanics are those of mathematics, it follows that mechanics is subalternated to mathematics. To all this that has been said one can add the testimony of Proclus,[21] a most weighty author, who set mechanics among the [S, 298r] parts of mathematics.[22] Further, mechanics has the characteristics that pertain to a science, so we shall say that it is a science. Most important, its subject has accidents that are not demonstrated by another science, which the mechanic demonstrates of it by means of its proper principles, and thus it is a science.

<II The Subject of Mechanics>

AN. And what is the subject of mechanics?[23]

PR. The subject of a science has to be such that its proper attributes come under consideration, as for example the subject of astronomy is the numerable movement of the heavens, since astronomy considers the properties of this movement. For this reason I say that the subject of mechanics is the machine, whence it comes to be called mechanics.

AN. And how will you prove it to be the machine?

PR. I shall prove it as follows. You remember having read in Plutarch and in Livy about the siege of Syracuse and about what Archimedes had done for Hiero, King of Syracuse? And what he did with one of his ships according to Plutarch?[24]

AN. I remember, but not well.

PR. Proclus narrates that Hiero had a huge ship built to send to Ptolemy, King of Egypt. And once built, that he wanted to launch it into the water, but the entire population of Syracuse was not sufficient to do it. At last Archimedes said to the king that he wanted him to launch it by himself. The king laughed at this. Archimedes made the machine and attached it to the ship, and had the king set the machine in motion, upon which it began to draw the ship into the water, so that before the king

fece che 'l Re desse il movimento alla machina, con il quale ella comincio à pingere la nave in acqua, la dove prima che 'l Re si partisse di quel luogo la nave fu tutta in acqua. Ciò vedendo il Re disse che da indi in poi crederebbe tutto quello che Archimede dicesse. Da quel che s' è detto, Vostra Signoria ha sentito l' effetto, et perché ciascun effetto è prodotto dalla sua cagione, però vedremo qual fu la cagione di far andar la nave in acqua. Et troveremo esser stata la machina, non come fatta di legni et di ferro, perché s' ha da presuporre che coloro e' quali tentarono di mandarvela havessero adoperato et legni et ferramenti per farlo, ma non successe loro. Adonque [S, 298v] diremo esser stata la virtù della machina, la quale altro non è che la proprietà di quella, et però siegue la machina essere il soggetto poi che di quella ne cerchiamo noi la proprietà. Così ancora, quando cercheremo le cagione perché le taglie fanno ó nel tirare i pesi ó nell' alzare quelli tanta forza, le taglie vengono ad essere il soggetto, et le proprietà di quelle vengono ad essere le cose che cerchiamo; le taglie adonque vengono ad esser ancor loro machina. Quel c' ho detto delle taglie dico dell' arteglierie, de gli horologii da ruote, et di tutte le cose che sono sottoposte al mecanico, come i molini di tutte le sorti, le machine da tirar acque, gl' instrumenti tutti da far forze, et così del resto.

AN. Resto capace che la machina sia il soggetto del mecanico, et ho letto appresso à Vitruvio che la diffinisce così: «La machina è una perpetua et continua congiontione di materia, che ha forza grandissima à movimenti de pesi». Ma non so vedere come egli faccia differenza tra machina et organo.

PR. La differenza è manifesta appresso à quello poi che gli la distingue dal lavoro. Diremo noi l' horologio da ruote essere machina, ma la stanga con la quale moveremo i pesi diremo organo ò stromento. In quello è molto lavoro et molte ruote; in questo puoco et niuna ruota. Ma è da vedere se veramente tra loro è differenza, perché il più lavoro et il meno lavoro non fa sensibile differenza ne varia la specie. Ma si bene il machinamento, ò l' havervi l' artefice pensato su, et però da tal machinamento chiamerò con ragione tanto machina il torchio con il quale si torchia il vino quanto una balestra ò arcobugio. Ma ciò non vuol dir nulla: concediamo à Vitruvio il chiamar quello col nome di «machina» è questo col nome di «stromento» et torniamo al primo proponimento.

Habbiamo adonque che 'l soggetto del mecanico è la machina, della quale se ne cercano le proprietà, le quale nascono dalla forma della machina, la quale è l' istessa [S, 299r] che 'l principio ò concorre al principio della machina, et però inteso il principio sapremo, et la forma et la

had left the place the ship was entirely in water. Seeing this the king said that henceforth he would believe everything that Archimedes said.[25] From what has been said you have heard the effect, and because each effect is produced by its cause, we shall determine what the cause was that made the ship go into the water. And we shall find it to have been the machine, not as made of ropes and iron, for one must suppose that those who tried to launch it would have used both ropes and ironworks to do it, but they did not succeed. Therefore [S, 298v] we shall say that it was the power of the machine, which is none other than its property, and thus it follows that the machine is the subject of which we now seek the property. So again, when we seek the cause why pulleys, either in pulling weights or in raising them, exert such force, the pulleys come to be the subject, and their properties come to be the things that we seek; so again, the pulleys come to be the machine. What I have said of pulleys I also say of artillery, of mechanical clocks, and of all the things that fall to the mechanic, such as mills of all sorts, machines for drawing water, all instruments to make forces, and so on.

AN. I can see that the machine is the subject of mechanics, and I have read in Vitruvius that he defines it thus: 'the machine is a perpetual and continuous construct of material that has the greatest force for the movement of weights.'[26] But I do not understand his distinction between machine and instrument.[27]

PR. The difference is clear because the one is distinguished from the other by workmanship. We would say that the mechanical clock is a machine, but the lever with which we move weights we would call an instrument. In the clock there is much workmanship and many gears; in the lever little workmanship and no gears. But it remains to be seen if there truly is a difference between them, because more or less workmanship does not make a sensible difference or change the species. But if the contrivance is good or if the craftsman has thought it out well, then from such contrivance I shall with as much reason call the press with which one presses wine a machine as a crossbow or arquebus. But this is not to say nothing: we concede to Vitruvius to call the first by the name of 'machine' and the second by the name of 'instrument' and return to the original purpose.[28]

We have agreed therefore that the subject of the mechanic is the machine, of which one seeks the properties, which arise from the form

cagione della virtù della machina. Ma non intendo per hora per ferma la figura di quella, ma la intrinseca virtù della machina.

AN. Mi nasce intorno alle cose dette un dubbio, ch' è questo. L' Altezza Vostra dal veder la machina et dal considerare le[13] proprietà di quella, è venuto in cognitione la machina essere il soggetto del mecanico, et però le fu facile trovare il soggetto dell' arte. Ma quei primi, che non hebbero la machina presente, come vennero in cognitione quella essere il soggetto della mecanica?

PR. Il dubbio è di facile resolutione, se Vostra Signoria si ricorda quel ch' habbiamo già detto poco fa. Fu detto che 'l fine del mecanico era di superare le difficultà con poche forze, il che non si può fare senza alcun mezo; quel mezo che viene operato in loco delle molte forze viene ad essere il soggetto. Come à dire, è il proposito di tirare un gran peso per terra, à ciò fare non bastano molti et molti huomini; trovasi che con l' applicarvi le taglie vien tirato da due ò da tre soli huomini. Adonque le taglie sono state mezo di far quello con poche forze che non si poteva fare con molte. Ma pare che due pezi di ligno con quelle girelle non havessero potuto far questo senza essere in esse alcuna occulta virtù; adonque le taglie hanno fatte quella operatione col mezo della virtù loro. Si cerca adonque la virtù delle[14] taglie, et le taglie tra tanto sono il soggetto. Se Vostra Signoria di nuovo ricercasse come[15] da principio furon ritrovate verbi gratia le taglie, direi da principio esser state ritrovate imperfettamente, perché come Vostra Signoria sa benissimo niuna cosa da principio fu trovata perfettamente, ma col tempo. Et però bene hanno detto i filosofi, che le arti et le scienze si riducono à perfettione per aggiugnimento. L' arteglierie da principio furon fatte rozamente et non si potevano maneggiare se non con molta difficoltà; et nel battere le mure non facevano effetto di momento se non con molto [S, 299v] et molto tempo. Gli arcobugi similmente, oltre l' imperfettione della forma, erano difficili in modo da maneggiare che nell' adoperarne uno bisognariano due et tre huomini. Ora Vostra Signoria può vedere à quanta perfettione sia venuta l' arte di tali strumenti. Haverà alcuno

13 le *ex* la *corr. S*
14 *ante* delle *del. S* loro
15 come] como *S*

of the machine, which is the same [S, 299r] as the principle or contributes to the principle of the machine, and so when we know the principle we shall know both the form and the cause of the power of the machine. But I do not intend to dwell now on the shape of the machine, but on its intrinsic power.

AN. There arises in me a doubt about these things, which is this. From seeing the machine and from considering its properties you concluded that the machine was the subject of the mechanic, and so it was easy for you to find the subject of the art. But those first mechanics, who did not have a machine at hand, how did they conclude that this was the subject of mechanics?

PR. The doubt is of easy resolution, if you remember what I said just a moment ago. I said that the goal of the mechanic was to overcome with small forces difficulties, which one cannot do without some means; the means that are used in place of great forces are the subject. For example, when the purpose is to draw a large weight over the ground, for which a very great many men are not enough, one finds that by the application of pulleys it can be drawn by two or three men alone. The pulleys, therefore, were the means to do with small forces what could not be done with large. But it would seem that two pieces of wood with these little wheels could not do it without being in essence some hidden power; therefore the pulleys did this work by means of their power. One seeks, therefore, the power of the pulleys, and the pulleys are its subject. Again, if you were to ask how pulleys, for example, were first discovered, I would say that they were first discovered imperfectly, because as you very well know nothing is discovered perfectly at first, but with time. And thus the philosophers spoke well, that the arts and the sciences are drawn to perfection through accretion. Artillery at first was made crudely and could not be managed except with great difficulty; and in battering walls it did not make any notable effect except over a very long [S, 299v] time. The arquebuses similarly, besides the imperfection of their design, were so difficult to manage that the operation of one of them needed two or three men. Now you can see to what perfection the art of such instruments has come. At first someone saw that pulleys with only one wheel made an effect of some moment, and so had one made of two and three wheels. Whoever first discovered the pulley

visto da principio che le taglie d' una sola girella haverà fatto effetto di alcun momento, et così n' haverà fatte di due et di tre. Chi trovo prima la taglia s' ha da credere che la trovasse per far più facilmente correre la corda col mezo della girella; visto poi che non solamente faceva al corso della corda, ma allegeriva la fattica in muovere il peso. Comincio à specularvi su in modo che vi venne trovando la cagione, et insieme speculando trovò il modo di multiplicarle la forza. La necessità, come da principio fa detto, porge molte volte occasione di trovare molte cose. Il caso ancora da materia à molte et à molte inventioni di non picciolo momento, si come forse è avvenuto alla cosa della polvere et dall' arteglierie, et si come è avvenuto à molte cose della medicina et dell' altre arti.

<III Il principio della meccanica>

<Le proprietà meravagliose del cerchio>

AN. Ho udito et inteso benissimo tutto quello che l' Altezza Vostra ha detto fin qui. Percioche ha detto che la proprietà della machina nasce dalla forma di quella ò da uno intrinseco principio suo, il quale è differente dalla figura della machina, chi adonque vorrà intendere le proprietà della machina sarà forza che intenda bene il principio suo. Adonque resta che l'Altezza Vostra si degni dichiararmi questo principio.

PR. Delle machine, si come ne sono di molte sorti, così ancora di quelle sono diversi principii. Ma per non confonderci dichiarerò prima il principio del quale parla Aristotele, et insieme delle machine delle quali habbiamo dato gli essempi, cioè delle taglie et della stanga, et poi discenderemo al resto. Vitruvio, come Vostra Signoria sa, nel luogo citato da lei, mette che 'l principio delle machinationi sia stato mostrato dalla natura al mecanico col movimento continuo del cielo; ma noi possiamo giugnere il movimento ancora delle cose et gravi et leggieri. [S, 300r] È adunque il principio communi delle cose mecaniche il movimento il quale primieramente con l' essere più veloce et più tardo, maggiore et minore effetto. Tutte le operationi delle machine adunque consistono nel movimento loro, et per consequente l' istessa machina farà maggiore et minore effetto quanto più propinquo sarà il movimento che se le farà fare al suo proprio. Poi che ciascuna machina consiste[16] in un certo temperamento di movimento, dal quale ciascuna volta, che ó per il troppo ó per il poco s' allontanasse il movimento che se le facesse fare, si corrom-

16 consiste] cosiste *S*

must have believed that he discovered it in order to make the passage of the rope easier by means of the wheel; then he saw that not only did it make the passage of the rope easier, but it lightened the effort of moving the weight. I begin to speculate about it in the way that leads me to find the cause, and while speculating I find the way to multiply the force. Necessity, as was said from the beginning, often presents occasions to find many things. Occasion again gives material to many people and for many inventions of no little moment,[29] as perhaps happened in the case of gunpowder and artillery, and as happened many times in medicine and the other arts.

<III The Principle of Mechanics>

<The Marvellous Properties of the Circle>

AN. I have attended to and understood very well all that you have said up till now. Since you said that the properties of a machine arise from its form or from its intrinsic principle, which is different from the shape of the machine, then whoever understands the properties of the machine would necessarily understand its principle. It remains, therefore, that you deign to explain to me this principle.

PR. Just as there are many sorts of machines, so also are their principles diverse. But in order not to confuse us I shall first discuss the principle that Aristotle states for both the machines we have given as examples, that is, for the pulley and the lever, and then we shall go on to the rest. Vitruvius, as you know, in the passage just cited by you, asserted that the principle of machination was shown by nature to the mechanic by the continuous movement of the heavens;[30] but we can also add the movement of both heavy and light things. [S, 300r] Therefore the principle common to mechanical things is the movement that is primarily and essentially swifter and slower, of greater and lesser effect. All the operations of machines, then, consist in their movement, and consequently the same machine will have a greater or lesser effect to the extent that the movement that it makes is nearer to its own, proper movement. For each machine consists of a certain arrangement of movement, by which as the movement that it produces gets farther away, whether by a lot or by a little, the force of the machine is reduced. And

perebbe la forza della machina. Et perché col mezo delle machine si fanno effetti che porgono maraviglia et che sono stimati miracoli, però è forza dice Aristotele che nascano da un principio miracoloso; ma nascono dal cerchio; adonque forza è nel cerchio essere molta maraviglia ò molte cose di maraviglia. Che nascono dal cerchio gli effetti di tutte ò della maggior parte delle machine è chiaro da questo: percioche si riducono le forze loro alla stanga, et questa si riduce alle libra, et questa finalmente al cerchio. Et però disse Archimede ad Hierone, «se mi fosse dato un luogo dove potesse appoggiare la stanga, io mandarei questa terra fuori del luogo suo». La dove appare che Archimede faceva le sue forze col mezo della stanga.

Che 'l cerchio ritenga in se cose maravigliose et sia miracolo, verremo dimostrandolo pian piano. Il cerchio nel nascimento suo ò nella sua fabrica è composto da due cose contrarie et però è miracolo, poiché non possono concorrere unitamente due contrarii à far alcuno effetto senza un mezo che l' unisca; ma nel cerchio concorrono senza tal mezo et però è miracolo. Intendono i matematici che 'l cerchio si descriva da[17] una linea retta che vadi intorno l' uno de gli estremi della quale sia fermo et per[18] centro, et l' altra descriverà la circonferenza di quello. Et però viene il cerchio ad essere descritto dal movimento et dalla quiete ò dall' immoto, et per ciò da due contrarii in uno stesso tempo senza mezo[19]. Appresso la circonferenza del cerchio, [S, 300v] essendo una linea senza largheza, ritiene in se due cose quasi contrarii, percioche ritiene il concavo et il convesso, che sono in un certo modo contrarii, perché il mezo loro è il retto: ne si può del concavo far convesso se prima quel che ha da passare dall' uno all' altro non sia fatto retto. Si come di due cose inequali: non può la maggiore passare nella minore ò questa nella maggiore, se non si passa per l' equale. Adunque diremo il cerchio esser fatto da due cose contrarii, et ritenere in se la circonferenza sua cose contrarie. Appresso il cerchio si muove con due movimenti contrarii, percioche una delle sue medietà si muove ascendendo et l' altra descendendo; ma l' insù et l' ingiù sono due luoghi contrarii; adunque il cerchio si muove con due movimenti contrarii, essendo la stessa cosa, il che è maraviglia non picciola. Questa conditione del cerchio è facile ad intendersi col movimento della bilanza, percioche mentre che un braccio di quella descende tirato all' ingiù dal peso, l' altro ascende, et così stante lo sparte ò il centro fermi, vengono le braccia à descrivere il cerchio.

17 da *inter lineas S*
18 per *corr. ex* poco *S*
19 mezo] moto *S*

because by means of machines one makes effects that produce wonder and are thought miracles, it is necessary, Aristotle says, that they arise from a marvellous principle; but they arise from the circle; therefore there must be in the circle many marvels or many sources of wonder.[31] That the effects of all or of most machines arise from the circle is clear from this: because their powers are reduced to the lever, and the lever is reduced to the balance, and the balance finally to the circle. And thus Archimedes said to Hiero, 'if I were given a place where I could put a lever, I would move this earth out of its place.'[32] Whence it appears that Archimedes produced his forces by means of the lever.

That the circle contains marvellous things and is a miracle, we shall see by proving it step by step. The circle in its origin or structure is composed of two contrary things and thus is a miracle, because two contrary things cannot combine into one. to produce an effect without some mean that unites them; but in the circle they unite without such a mean and thus it is a miracle. Mathematicians say that the circle is described by a straight line that goes around one of the ends, which is fixed and at the centre, and the other end describes its circumference. And so the circle comes to be described by motion and by rest or the unmoved, and thus by two simultaneous contraries without a mean. Further, the circumference of the circle, [S, 300v] being a line without width, contains in itself two almost contrary things, for it contains the concave and the convex, which are in a certain way contraries, because their mean is the straight: one cannot make the convex from the concave unless first what is to pass from from the one to the other is made straight. Just as of two unequal things the larger cannot pass into the smaller, or the smaller into the larger, unless it passes through the equal. Therefore we say that the circle is made from two contrary things and contains in its circumference contrary things. Further, the circle moves with two contrary movements, since one half ascends and the other descends; but up and down are two contrary places; thus the circle, although one and the same thing, moves with two contrary movements, which is no small marvel. This property of the circle is easy to accord with the movement of the balance, because while one arm, being drawn downward by the weight, descends, the other ascends; and thus, with the pivot or centre standing fixed, the arms describe a circle.[33]

<Il principio del movimento circolare>

Ma queste proprietà della figura circolare et queste dignità et maraviglie non fanno ponto al caso nostro; anzi molte altre sono le proprietà maravigliose della figura circolare le quali noi le pretermittiamo, come cose che non fanno al proposito nostro. Solo fa al proposito del mecanico quello che hora diremo del cerchio. È una proprietà nel cerchio dalla quale nascono le forze mecaniche che si possono à quello ridurre. La qual proprietà è, ch' essendo una linea sola quella che descrive il cerchio, nondimeno i ponti presi su quella non si muove con equal velocità, ma quei che più appresso <al centro> sono si muovono più tardamente di quei che più da quello se ne allontanano. Il tutto si può considerare dall' essempio de cieli, poiché, considerato il movimento loro da levante in ponente, viene il movimento del primo mobile ad esser velocissimo, et quello della luna tardissimo. Et la cagione è che quello del primo mobile passa in 24 hore una grandeza immensa [S, 301r] ch' è la circonferenza dell' equinotiale suo, ma quel della luna nello stesso tempo passa picciolissimo spatio, rispetto à quello del primo mobile. Come Vostra Signoria sa, poi il movimento si dice tardo et veloce così rispetto allo spacio come rispetto al tempo. Percioche diremo noi un mobile essersi mosso con maggior velocità d' un' altro quando in uguale tempo haverà passato maggiore spacio, come nell' essempio proposto. Così ancora diremo un mobile esser più veloce d' un' altro quando ūguale spacio l' haverà passato in minor tempo. Come se due andasserò per uno stesso camino à Roma, ma che l' uno v' andasse in quattro di et l' altro l' uno et mezo, ó poco più ó meno, dove appare uno stesso spatio essere da detti passato in tempi disuguali. Più veloce adunque diremo essersi mosso il secondo che il primo.

AN. Et quando due motori passano in tempi equali spacii equali, i movimenti sarano adonque equali. Ma questa cosa della linea che descrive il cerchio del non moversi con egual velocità tuttavolta è[20] in molte cose che l' ho osservato: come nelle pietre che aguzano i cortelli, nelle ruote con le quali si bruniscono le armi. Si può vedere nelle pietre de molini, nelle quali appare che quanto più i punti s' allottano al centro, tanto più tardamente si muovono, anzi quei ponti che sono propinquissimi al centro, perché non si muovano. Ma che cosa vuol da ciò conchiudere Aristotele? – perche io non so vedere à che proposito possa dirlo.

PR. Vostra Signoria stia ad udire che vedrà di quanto momento è tal

20 tuttavolta è] tutta oltre *S*

<*The Principle of Circular Movement*>

But these properties and these attributes and marvels of the circular figure do not pertain to our case; in fact, there are many other marvellous properties of the circular figure that we shall omit, as things that do not advance our purpose. Only the one property of the circle that we shall now state concerns the mechanic. It is a property of the circle from which arise the mechanical forces that can be reduced to it. This property is that, although a circle is described only by a line, nevertheless the points taken on it do not move with equal speeds, but those that are nearer <the centre> move more slowly than those that are farther from it.[34] One can see it all in the example of the heavens, for, considering the movement from east to west, the movement of the *primum mobile* tends to be the swiftest and that of the moon the slowest. And the reason is that in twenty-four hours the movement of the *primum mobile* traverses an immense distance, [S, 301r] which is the circumference of the celestial equator, but in the same time that of the moon traverses a very small distance in comparison to that of the *primum mobile*. As you know, movement is called slow and swift with respect to space as well as with respect to time. For this reason we say that a mobile is moved with a greater speed than another when in an equal time it has traversed a greater distance, as in the example given.[35] So again we say that a mobile is swifter than another when it has traversed an equal distance in less time. As <for example> if two were to go by the same road to Rome, but one gets there in four days and the other in a day and a half, more or less, where the same distance is traversed by them in unequal times. We say then that the second is moved more swiftly than the first.

AN. And when two movers traverse equal spaces in equal times, the movements will thus be equal. But this effect of the line that describes a circle of not being moved with equal speed overall <is apparent> in many things that I have observed: as in stones that level courtyards, in wheels with which arms are burnished. One can see it in millstones, where it appears that the closer the points are to the centre, the more slowly they move; in fact, those points that are closest to the centre hardly move at all. But what does Aristotle want to conclude from this? – because I do not see what his purpose is for saying it.

PR. Listen a moment so that you will be able to see of how much

principio, ma non così semplicemente preso. Ha havuto nell' intentione Aristotele di ritrovar la cagione perché una stessa virtù più appresso al centro facci meno forza che la stessa più da quello lontana, si come appare nella statera, nella quale quanto più il nomano ó il marchio ó quel peso che più et meno s' accosta ò discosta dal centro fa contrapeso à più et à meno peso. À maggior peso contrapesa quanto più dal centro s' allontana et à minore quanto più à quello s' accosta. Et centro chiamo quella parte dell' asta [S, 301v] dove è sospesa et si volge l' asta all' insù et all' ingiù. Volendo adunque Aristotele trovar la cagione di questo, ricorre al cerchio et al modo con il quale quello si descrive, et dimostra che quanto più un ponto s' allontana dal centro nella linea che quello descrive, con tanto maggior velocità si move; et poco conseguente una stessa virtù più lontana dal centro applicata, aiutata dalla velocità naturale del cerchio, tanto maggior forza farà. Or per dimostrare che quanto più il punto s' allontana dal centro con tanto maggior velocità si muova: presuppone la linea che descrive il cerchio moversi con due movimenti, l' uno de quali chiama secondo la natura della linea et l' altro contra la natura di quella. Et fa ancora un presupposto che tai due movimenti sieno disgiunti d' ogni proportione in ogni tempo; perché quando due movimenti sono congionti in alcuna proportione per alcun tempo, vengono à descrivere una linea retta.

AN. Et quali sono i due movimenti della linea che descrivono il cerchio?

PR. Sono i seguenti. Intendiamo il cerchio ABCD descritto sopra il centro E, et sopra l' istesso centro E intendiamovi descritto un' altro cerchio minore HIKL [*vide figuram I*]. Et sia una linea che venghi dal centro E et passi per ambe le circonferenze EIB.[21] Chiara cosa è, che quando la linea EIB[22] havesse da moversi secondo il suo naturale movimento, che non in giro ma per dritto si moverebbe, et verrebbe à descrivere un parallellogramo rettangolo. Ma perché viene ad essere obligata al centro, però viene a descrive cerchio. Et perciò nel proposto essempio della linea EIB[23] dovendosi ella movere secondo il suo naturale [S, 302r] movimento verrebbe à descrivere portata su l' altro semidiametro del cerchio la figura parallellograma EBMC. Ma mettiamo, che non habbi passato tutta la linea EC ma solamente la parte di quella EG; haverà all' hora descritto il

21 EIB] ELB *S*
22 EIB] ELB *S*
23 EIB] ELB *S*

importance this principle is, though it is not so easily grasped. Aristotle wanted to find the reason why the same power nearer the centre produces less force than farther from it, as appears in the steelyard, in which the counterweight or the pile driver or some weight counterbalances more or less weight as it is put closer to or farther from the centre. It counterbalances a greater weight when it is farther from the centre and a lesser when it is closer. And we call that part of the arm the centre [S, 301v] where the arm is suspended and pivots up and down. Wanting therefore to find the cause of this, Aristotle had recourse to the circle and to the way by which it is produced, and he showed that the farther a point is from the centre on the line that describes it, the more speed it is moved with; and so the same power applied farther from the centre, aided by the natural speed of the circle, will produce so much more force. Now, to demonstrate that how much farther the point is from the centre, with so much more speed it moves: suppose the line that describes the circle moves with two movements, one we call according to the nature of the line and the other against its nature. And again, suppose that two such movements are devoid of any ratio to each other in any time; because when two movements are related in some <fixed> ratio through some time, they come to describe a straight line.[36]

AN. And what are the two movements of the line that describe the circle?

PR. They are the following. Let the circle ABCD be described around centre E, and around the same centre E let another, smaller, circle, HIKL, be described (see figure I). And let there be a line EIB that comes from the centre E and intersects both the circumferences. It is clear that if the line EIB were moved according to its natural movement it would be moved not in a circle but in a straight line, and it would describe a rectangular parallelogram. But because it happens to be constrained at the centre, it tends to describe a circle. And thus in the proposed example of the line EIB, having to move according to its natural [S, 302r] movement, it would describe the parallelogram EBMC placed on a radius of the circle. But let us suppose that it does not traverse all the line EC but only a part of it EG; it would now have described the parallelogram EBFG, one side of which would be EIB and the other GNF. This posited, I say that point I would be moved more against its

parallellogramo EBFG, l' uno de lati del quale sarà EIB[24] et l' altro GNF. Ciò stante, dico che 'l ponto I si sarà mosso più contra il natural movimento suo, che non haverà fatto il ponto B. Perché il movimento naturale della linea sarebbe per le rette[25] BF et EG, il che non potendo essere per la violenza che 'l ponto E fa con l' essere in quel luogo fermo; però il ponto B si moverà per il natural suo per la circonferenza BR et il ponto I per la circonferenza IN, et tal movimento chiameremo con Aristotele[26] movimento secondo la natura della linea. Et perché dovendo il punto B descrivere la circonferenza BR, vien rimosso dal suo movimento retto, che sarebbe BF, per lo spacio FR. Perciò lo spacio FR si dirà movimento violento ò ritirato della linea, et così ancora lo spacio NO.

Da questi presupositi, è facile provare poi lo spacio FR (ò BQ[27] à quello equale) essere minore dello spacio ON (equale ad IP). Da dove seguirà ch' essendo il ponto I più violento del ponto B ò più di quello ritirato, che però meno contra la natura sua si moverà che non farà il ponto I[28]. Essendo chiara cosa che di due cose che venghir pinti et che sieno d' uguale forza, che quel sarà meno pinto che più resiste, et quello più che meno; o di due cose d' ugual forza che si muovano, quello si moverà

24 EIB] ELB *S*
25 le rette] il retto *S*
26 Aristotele] Aristole *S, ante quod scrip. et del.* Aristotele *S*
27 BQ] BQi *S*
28 I] B *S*

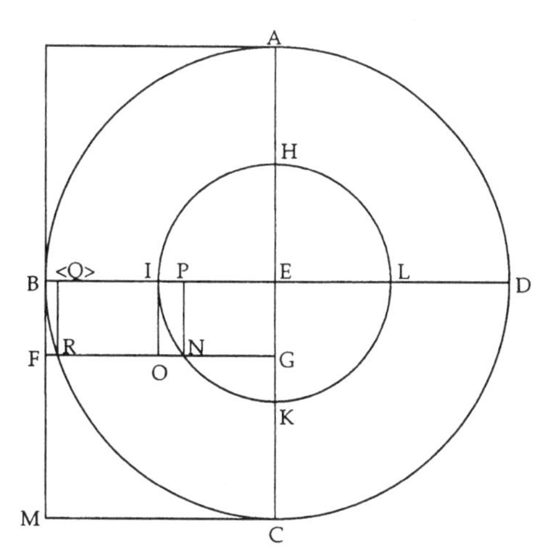

Figure I

natural movement than point B would be. Because the natural movement of the line would be along the straight lines BF and EG, which cannot be because of the violence that point E produces by being in a fixed place; but point B would be moved by its natural <movement> along the arc BR and point I along the arc IN and such movement we shall call with Aristotle the natural movement of the line. And because point B must describe the arc BR, it is deflected from its straight movement, which would be BF, by the distance FR. For this reason the distance FR is called violent movement or movement deflected from the line, and similarly the distance NO.

From these presuppositions, it is easy to prove that the distance FR (or BQ, which is equal to it), is less than the distance ON (or IP). From this it follows that, since point I is more violent than point B or more deflected than it, point B would thus be moved less against its nature than point I would be. Now it is clear that of two things that are pushed and are of equal force, the one that resists more will be pushed less, and vice versa; or of two things of equal force that are moved, the one that is more restrained by another force will be moved less, and the one that is less restrained more. If this were not self-evident, one could prove it by

meno che più d' un' altra forza sarà ritenuto, et quello più che meno ritenuto sarà. Il che se non fosse manifesto da se, si potrebbe dimostrare col mezo d' alcuni principii. Ciascuna volta adunque che s' havrà provato che lo spacio BQ sia minore dello spacio IP, s' haverà l' intento.

AN. Bel pensiero è stato veramente quello d' Aristotele et nobile imaginatione, ma [S, 302v] venghiamo alla demostratione.

<Proposizione I. Che un cordo nel circolo minore più s' allontani dalla circonferenze che un cordo eguale nel circolo maggiore>

PR. Potrà io di questa cosa farne più dimostrationi ò potrei addurne diverse dimostrationi, ma porterò à Vostra Signoria la più facile et che pur hora ho ritrovata. Quello adonque che s' ha da dimostrare e questo. Se in due cerchi inequali s' applicheramo due linee equali, il pezo della perpendicolare che verrà dal centro sopra l' una et l' altra ne cerchi et che si distenda fino alla circonferenza, che sarà tra la circonferenza et la linea nel cerchio maggiore sarà minore, che 'l pezo tra la linea et la circonferenza nel minore. Sieno adonque i cerchi inequali, ABCD maggiore et FGIH minore, sopra il centro E, et sieno in essi le due linee equali, BD nel maggiore et GH nel minore, la qual GH ha da essere minore del diametro FEI, perché quando fosse à quello equale, sarebbe manifesto quel che si vuole [*vide figuram II*]. Dal centro E ceda la linea perpendiculare sopra et BD et GH et distendendosi nelle circonferenze ne ponti A et F. Dico la portione del semidiametro del cerchio minore <essere maggiore>.

means of some principles. Any time, then, that length BQ is proved to be less than length IP, the purpose is accomplished.[37]

AN. That was truly an elegant and noble conception of Aristotle's, but [S, 302v] let's come to the demonstration.

<Proposition I. That a chord in a smaller circle is farther from the circumference than an equal chord in a larger circle>

PR. I could make more demonstrations of this or I could produce different demonstrations, but I shall present to you the easiest I have found as yet. What is to be demonstrated, then, is this. If in two unequal circles we put two equal chords, of the perpendicular that goes from the centre to each of the chords in the circles and that extends to the circumference, the part between the circumference and the chord in the larger circle will be smaller than the part between the chord and the circumference in the smaller. Therefore let there be two unequal circles, ABCD the larger and FGIH the smaller, around the centre E, and let there be in them two equal chords, BD in the larger and GH in the smaller (GH must be less than the diameter FEI because if it were equal to it, what one desires <to prove> would be clear) (see figure II). From the centre E the perpendicu-

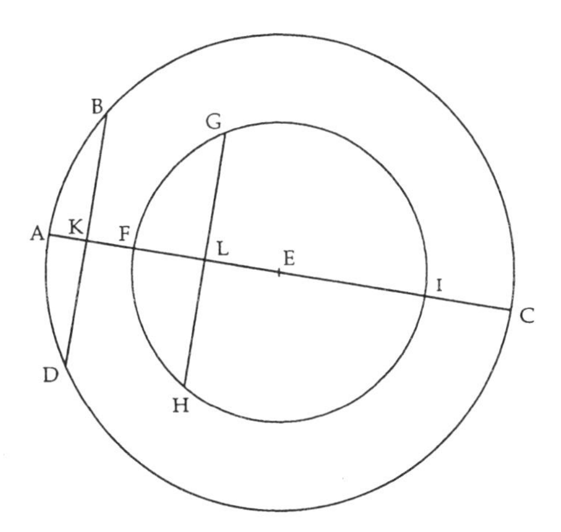

Figure II

lar line intersects both BD and GH and is extended to the circumferences at points A and F. I say that the part of the radius of the smaller circle <is greater>.

Di nuovo, descrivasi il cerchio ABCD, maggiore ò equale al maggiore, sopra il centro E, et tirisi il diametro AEC [*vide figuram III*]. Et dal ponto A verso il ponto E, piglisi la linea AF equale al semidiametro del cerchio minore [*in figura II*], et sopra il centro F et l' intervallo A<F> descrivasi il cerchio minore AGHI. Sopra il ponto E poi tirisi il diametro BED ad angoli [S, 303r] retti sopra AEC, et parimente sopra il ponto F tirisi il diametro GFI ad angoli retti sopra AH. Così del punto E verso B come verso B come verso D, pigliasi due pezi di linea equali ciascuni[29] alla mita della linea BD ò GH dell' altra figura [*i.e., figurae II*], et sieno EL et EK. Et sopra i ponti L et K alzirsi due perpendicolari et produchirsi fino alla circonferenza del cerchio ABCD ne ponti O, P. Et dal ponto O al ponto P tirisi la linea OSP. Non è dubbio le dette perpendicolari tagliare ancora la circonferenza del cerchio minore, essendo la linea KL minore del diametro HA ò GI. Il quale cerchio essendo dentro al maggiore, i ponti delle settioni saranno sotto i ponti O, P; sieno adunque Q R per i quali tirisi QTR. Chiara cosa è le linee OSP, QTR, et KEL essere per la 33ª et 34ª del primo d' Euclide et per la 14ª del terzo di quello equali tra di loro. Ma il ponto T è sotto il ponto S; adonque maggiore è TA di AS, ch' è quello che si voleva dimostrare.

AN. Venghiamo al resto se pare all' Altezza Vostra et à provare i due primi presupposti.

29 ciascuni] crescano_ *S*

Again, draw circle ABCD, larger than or equal to the larger <circle in figure II>, around centre E, and draw the diameter AEC (see figure III).

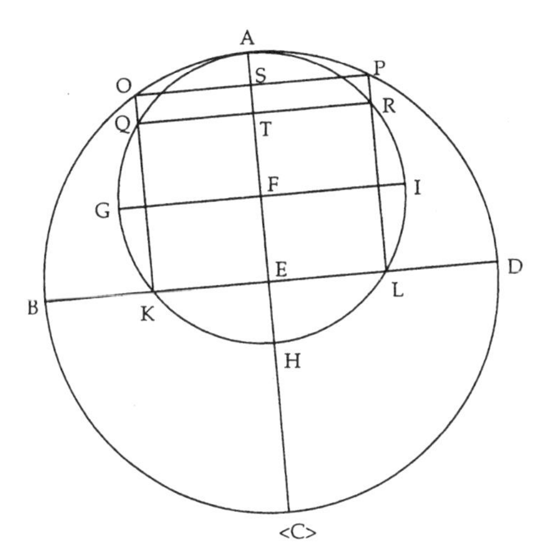

Figure III

From point A towards point E take line AF equal to the radius of the smaller circle <in figure II>, and around centre F and with radius A<F> draw a smaller circle AGHI. Through point E then draw the diameter BED at right [S, 303r] angles to AEC, and similarly through point F draw diameter GFI at right angles to AH. From point E towards both B and D, take two line segments each equal to half of line BD or GH of figure II, and they are EL and EK. On points L and K raise two perpendiculars and produce them to the circumference of the circle ABCD to points O and P. And from point O to point P draw line OSP. There is no doubt that these perpendiculars again cut the circumference of the smaller circle, since line KL is less than the diameter HA or GI. Since this circle is inside the greater, the points of intersection will be below points O and P; therefore they are Q and R, through which QTR is drawn. It is clear, by Euclid, *Elements* 1.33 and 34, and 3.14, that lines OSP, QTR, and KEL are all equal.[38] But point T is under point S; therefore TA is larger than AS, which is what was to be demonstrated.

AN. Let us please proceed to the rest and prove the first two suppositions.[39]

<Proposizione II. Che due movimenti retti in alcuna proporzione congionti descriveranno la linea retta>

PR. Il primo presupposto fu che ogni volta che due movimenti sono in alcuna proportione[30] congionti, che descriveranno una linea retta. Vostra Signoria ha da sapere che i movimenti, con i quali vien descritta la linea retta, ó sono retti, ó circolari, ó l' uno retto et l' altro circolare. Sieno prima retti et intendarsi così. Mettiamo due linee che si congionghino in angolo et sieno AB et AC, le quali sieno ó equali ó inequali, che non fa caso, et sieno le linee terminate da ponti B et C [*vide figuram IV*]. Et mettiamo che tra tanto che la linea AC si muove su la linea AB, descrivendo il parallellogrammo ABCD, ancora il ponto A si muova su la linea AC verso C. Dico, stante questo presupposto, che 'l ponto [S, 303v] A verrà descrivendo il diametro AD. Intendiamo hora che la linea AC si sia mossa su la linea AB per lo spacio AE. Il ponto E, tra tanto si sarà mosso per la linea AC, per tanto spacio che comparato allo spacio AE haverà à quello, quella proportione c' haverà AC ad AB, et così si sarà mosso per AF. Ma gia AC si sarà mossa su la linea AB; adonque il ponte F sarà nel ponto G, perché, essendo il parallellogrammo AEFG proportionale al parallellogrammo ABCD (per la 26ª del sesto d' Euclide), sarà intorno allo stesso diametro; et però il ponto G sarà sul diametro AD et haverà, col movimento et suo et portato tra tanto, passato lo spacio AG del diametro AD. Il simile può intendere Vostra Signo-

30 proportione] preparatione *S*

<Proposition II: That two rectilinear movements in any fixed ratio describe a straight line>

PR. The first supposition was that every time that two movements are related in any ratio they describe a straight line. You must know that movements, when they are to describe a straight line, are either both straight, or both circular, or one straight and the other circular. Let them first be straight and proceed as follows. Let us posit two lines that met at an angle and let them be AB and AC, which are either equal or unequal (it makes no difference), and let the lines be terminated by points B and C (see figure IV). Let us also posit that as line AC is moved along line

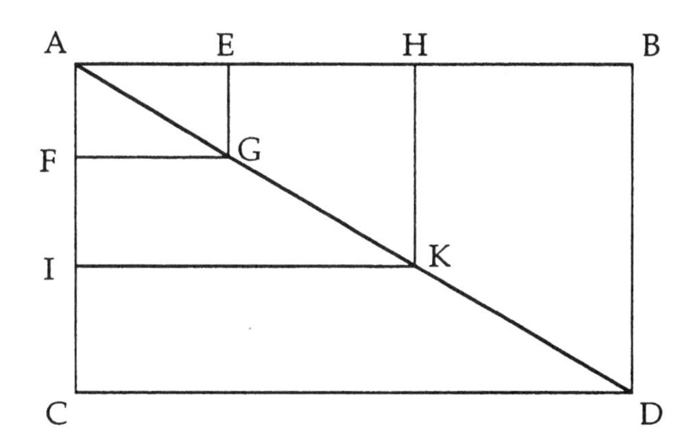

Figure IV

AB, describing the parallelogram ABCD, so the point A is moved along line AC towards C. This posited, I say that point [S, 303v] A will describe the diagonal AD. Let us now propose that line AC be moved along line AB through the distance AE. Point E would be moved along line AC through a distance that will have the same ratio to the distance AE that AC has to AB, and thus it would be moved through AF. But AC would already have been moved along line AB; therefore point F would be at point G, because the parallelogram AEFG, being similar to parallelogram ABCD (by Euclid, *Elements*, 6.26),[40] would be on the same diagonal, and thus point G would be on the diagonal AD and would have traversed, both with its own movement and with being carried along, the distance AG of diagonal AD. You can understand the same of parallelogram AHIK, where it appears that when line AC, travelling

ria del parallellogrammo AHIK, la dove appare che quando la linea AC, caminando su la linea AB, haverà descritto il parallellogrammo ABCD, che' l ponto A, caminando su la linea AC, haverà descritto il diametro AD, ch' e una retta linea, ch' e quello, che si voleva dimostrare. Se Vostra Signoria hora darà mente à questo stromento, la vedrà che fa mecanicamente l' istesso effetto.

AN. È vero et è bello, ma descendiamo a dimostrare come i due movimenti circolari descrivano la retta.

<Proposizione III. Che con due movimenti circolari possa descriversi la linea retta>

PR. Intenderemo un cerchio ABCD sopra il centro E[31], et nella circonferenza di quello intenderemo descritto un' altro cerchio ad esso equale sopra il centro A, che[32] sarà FGED[33], et produremo la linea retta FAECN [*vide figuram V*]. Appresso intenderemo il cerchio ABCD moversi sopra il suo centro E, portando seco l' altro cerchio attorno, et tra

31 E] AE *S*
32 che] che che *S*
33 FGED] FGHD *S*

along line AB, has described the parallelogram ABCD, point A, travelling along line AC, has described the diagonal AD, which is a straight line, which is what was to be demonstrated.[41] If you now will turn your attention to this instrument here, you will see that it produces mechanically the same effect.[42]

AN. It is true and elegant; but let us now proceed to demonstrate how two circular movements describe a straight line.

<Proposition III. That two circular movements can describe a straight line>

PR. Let us posit a circle ABCD around centre E, and on its circumference let us suppose another circle is drawn equal to it around centre A, which would be FGED; and let us draw the straight line FAECN (see figure V). Next, let us suppose circle ABCD is moved around its centre

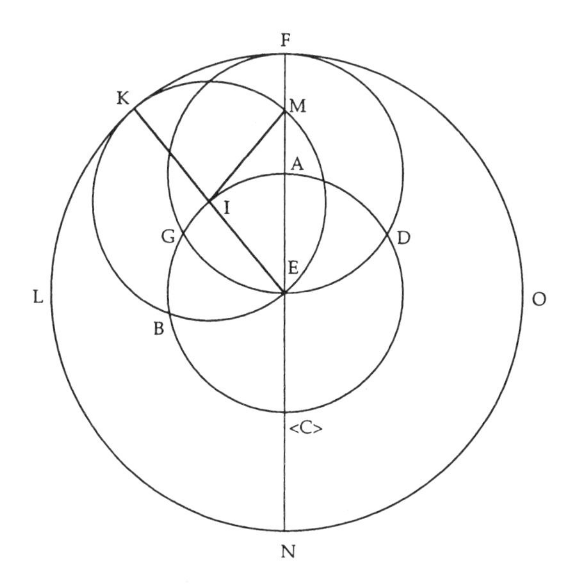

Figure V

E, carrying around with it the other circle, and let us suppose circle FGED is meanwhile moved around its centre. Now, I want point F of the

tanto intenderemo moversi il cerchio FGED sopra il suo centro. Volgio hora che 'l ponto F del cerchio FGED descriva la linea [S, 304r] retta FECN. Farò adonque che tra tanto che 'l cerchio ABCD si moverà col ponto A verso B, ò dalla sinistra, per una revolutione, che 'l cerchio FGED si doveva col ponto F per due revolutioni verso la destra[34] di modo che quando il ponto A haverà descritto una quarta del suo cerchio, il ponto F habbia passato una mita della sua circonferenza. Et però tra movimenti sarà proportione doppia. Or sia adonque il ponto A venuto nel punto I, et fatto I centro descrittovi sopra il cerchio KMB, quando il ponto F fosse immobile egli sarebbe nel punto K; ma perché habbiamo detto moversi verso la destra con doppio movimento, però se nel arco KME piglieremo l' arco KM doppio ad AI, haveremo il ponto F essere su la linea FAE nel ponto M. Et tra tanto che 'l ponto A sarà venuto in I, il ponto F haverà passato tutto il pezo FM della linea FAE. Et continuando i due cerchi i movimenti loro con l' istessa proportione, quando il ponto A sarà per una linea ad angoli retti sul ponto E della linea FN, il ponto F, havendo passato et descritto tutta la linea FAE, sarà venuto nel ponto E; et continuando ancora i movimenti loro, quando il A sarà venuto nell' opposito suo, cioè nel ponto C, il ponto F haverà descritto continuando tutta la linea ECN.

Che poi il ponto F sia venuto nel ponto M è chiaro, et si dimostra così. Dal presupposto, l' angolo KIM è doppio all' angolo IEM; ma l' angolo KIM (per la 32ª del primo d' Euclide) è equale all' angolo IEM et all' angolo IME; adonque l' angolo IME sarà equale all' angolo IEM; et però (per la sesta di quello) il lato IM sarà equale al lato IE; adonque il ponto M caderà su la linea FAE di necessità. Questa demostratione può servire ad ogn' altro ponto, perché quando il punto M non fosse su la linea FAM, l' arco KM non sarebbe doppio all' arco AI, contra quello che si suppose. Da qui ancora può vedere Vostra Signoria come la circonferenza FGD viene ad haver descritto [S, 304v] il cerchio FKNO. Ma è da sapere che non potrebbono farsi movere i cerchi in qual si voglia altra proportione. Se bene però i movimenti fossero contrarii et il movimento più lento ò tardo fosse fatto dal cerchio ABCD, perché non verrebbe l' istesso, percioche il ponto F non descriverebbe la linea retta FEN, ma sarebbe da quella lontano, ò di qua ò di la di FAE.

AN. Non è stata men bella questa imaginatione dell' altra, anzi tanto più di acuta inventione quanto è più lontano il movimento circolare dal movimento retto. Ma dicami di gratia l' Altezza Vostra, è di sua inventione?

34 destra] desta *S*

circle FGED to describe the straight [S, 304r] line FECN. I shall arrange it, then, so that while the circle ABCD is moved point A towards B, or to the left, through one revolution, the circle FGED must <move> point F to the right in such a way that when point A has described a quarter of its circle, point F will have traversed half of its circumference. Thus between the movements there will be a ratio of two to one. Now, when point A, therefore, comes to point I, and circle KMB is described around centre I, if point F were immobile it would be at point K; but because we have said that it is moved to the right with twice the movement, then if on the arc KME we choose the arc KM double AI, we would have point F be on line FAE at point M. And while point A has come to I, point F will have traversed all of the part FM of line FAE. And with the two circles continuing their movements in the same ratio, when point A is on a line through point E at right angles to line FN, point F, having traversed and described the entire line FAE, will have reached point E; and continuing their movements farther, when A has reached its opposite, that is, point C, point F will have described the entire line ECN.

That point F will come to point M is clear, and is demonstrated as follows. By supposition, angle KIM is twice angle IEM; but angle KIM (by Euclid, *Elements*, 1.32) is equal to angle IEM plus angle IME;[43] therefore angle IME will be equal to angle IEM; and thus (by Euclid, *Elements*, 1.6) side IM will be equal to side IE;[44] therefore point M will necessarily fall on line FAE. This demonstration can serve for every other point, for if point M were not on line FAM, arc KM would not be twice arc AI, contrary to what was posited. From this you can further see how circumference FGD comes to describe [S, 304v] the circle FKNO. But it must be understood that the circles cannot be made to move in any other ratio you like. For if the movements were contrary and the movement were made faster or slower by circle ABCD, then it would not produce the same result because point F would not describe the straight line FEN, but it would be off to one side of it, either on this side or on the other side of FAE.

AN. This was no less elegant a conception than the other, in fact, so much more subtle an invention as circular movement is different from rectilinear.[45] But tell me please, is this your own invention?

PR. Iddio volesse che fosse mia! È del Copernico, huomo dottissimo, nel suo libro *Della revolutione de cieli.*

AN.[35] È il Copernico quello che s' imagina la terra mobile per salvare l' apparente?

PR. È quello.

AN. Mi ricordo ch' un mio vassallo mi leggeva *La sfera* et Euclide, et tratto per tratto le dava addosso, adducendo molti inconvenienti contra di lui.

PR. Questo nasceva forse perché non intendeva il fine del Copernico; ma di questo un' altra volta se pare à Vostra Signoria.

AN. Come Vostra Altezza comanda. Haverà hora da mostrare come col movimento retto et circolare insieme possa descriversi la linea retta.

<Proposizione IV. Come col movimento retto et circolare insieme possa descriversi la linea retta>

PR. Mentre che udivo *Le mecaniche,* ritrovai da me il modo, il quale è questo. Intenderemo primieramente sopra il centro E descritto il cerchio ABCD, et sopra il centro E intenderemo cadere la perpendicolare AE sul diametro BED [*vide figuram VI*]. Intenderemo appresso il cerchio ABCD muoversi sul suo diametro per il dritto, cioè su la linea EB, et tra tanto

35 AN.] E AN. *S*

PR. Would to God it were mine! It is Copernicus's, a most learned man, in his book *On the Revolutions of the Heavens*.[46]

AN. Is this the Copernicus who imagines the earth is mobile to save the appearances?

PR. The same.

AN. I remember that a vassal of mine read me the *Sphere* and Euclid,[47] and passage for passage used them to attack him, adducing many arguments against him.

PR. This arose perhaps because he did not understand Copernicus's intent; but of this another time if you like.[48]

AN. As you command. Now you must show how with a straight and a circular movement together a straight line can be described.

<Proposition IV. How with a straight and circular movement together a straight line can be described>

PR. When I heard the *Mechanics* I found by myself the way, which is this.[49] First let us suppose that around centre E is described the circle ABCD, and through centre E let us suppose the perpendicular AE falls

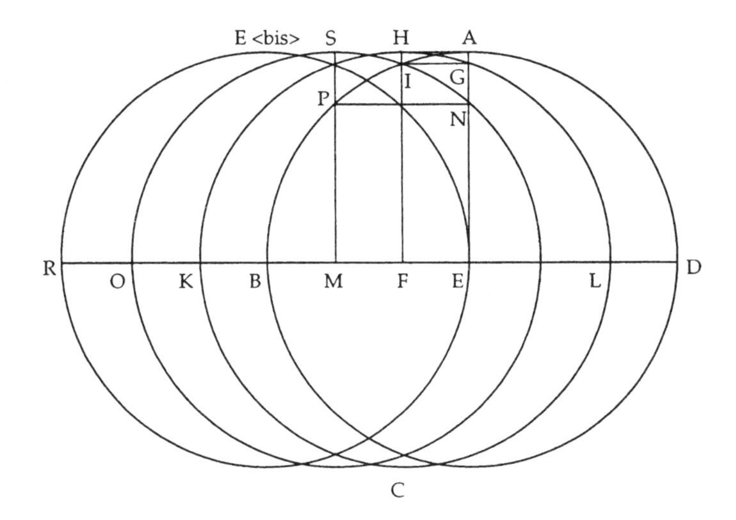

Figure VI

to the diameter BED (see figure VI). Next let us suppose that the circle ABCD is moved on its diameter along a straight line, that is along line

moversi in giro verso D. Or dico che mentre il centro E haverà [S, 305r] caminato il suo semidiametro, et che 'l ponto A haverà caminato una quarta del cerchio, il ponto A stesso haverà descritto una linea retta equale al semidiametro del cerchio suo.

Presupponghiamo adonque il centro E essersi mosso fino al ponto F, et intendiamo descritto sul centro E la quarta APB; et poi, fatto centro F et l' intervallo equale ad EB, descriviamo un cerchio equale al cerchio di prima et sia HKCL, et alziamo dal ponto F la perpendicolare su la linea BE et sia FH, la quale tagli la circonferenza APB nel ponto I, et dal ponto I tirisi una linea parallela alla linea EB et sia IG. Dico che nel ponto G sarà venuto il ponto A et il ponto G sarà in diretto col ponto A; così dico ancora ch' essendosi mosso il centro E per il detto movimento fino al ponto M, che 'l ponto A sarà venuto nel ponto N et sarà parimente in diretto col ponto A et G. Et quando poi il ponto E sarà venuto nel ponto B, il ponto A haverà passato tutti i ponti che sono tra AE et haverà descritto la linea AE.

Producasi la linea dal punto H al ponto A. Perché la linea FH è ad angoli retti su la linea BE et così la linea EA, adonque (per la 27ª del primo d' Euclide) saranno le due dette parallele, et HA sarà (per la 33ª del detto) equale et parallela all' FE. Ma parallela è IG[36] ad FE; adonque (per la 30ª del primo d' Euclide) sarà parallela all' AH et à quella equale, et però equale sarà l' arco AI all' arco HG. Adonque il ponto A haverà caminato l' arco HG et sarà pervenuto nel punto G, et N et G sono in diretto per le ragioni dette.

L' istesso proveremo del ponto P. Percioche essendo venuto il punto E nel punto M alzato dal punto M la perpendicolare MPS et tagliando quella la quarta nel punto P, dico il punto A esser venuto nel punto N. Perché quando il ponto A si fosse mosso per la quarta APB con la stessa proportione, sarebbe venuto nel ponto P; ma l' arco AP è equale all' arco SN, essendo che la [S, 305v] linea che si stende à tutte due è l' istessa. Però (per la 28ª del terzo et per la terza di quello), essendo i cerchi equali, s' haverà l' intento il simile si può provare dall' equalità de cerchi et dalla similitudine de movimenti. Habbiamo adonque con i due movimenti, l' uno retto et l' altro circolare, descritto la linea retta, ch' è quello che si voleva fare. Di tutto se ne possono fare instromenti ne quali benissimo si potrebbe vedere l' effetto.

AN. Per non dar più travaglio all' Altezza Vostra, mi contento per hora delle dimostrationi dette percioche l' ho capite benissimo. Può l' Altezza Vostra essendo servita passare al resto?

36 è IG] EIG *S*

EB, and at the same time it rotates towards D. Now, I say that when the centre E has [S, 305r] traversed its radius, and point A has traversed a quarter of the circle, point A itself will have described a straight line equal to the radius of its circle.[50]

Let us suppose, then, that the centre E has been moved to point F, and let us posit that the quarter-circle APB is described around the centre E; and then, with F as centre and a radius equal to EB, let us draw a circle HKCL equal to the first circle, and let us raise from point F a line FH perpendicular to line BE, which would cut the circumference APB at point I, and from point I draw line IG parallel to line EB. I say that A will have come to point G and point G will lie on a straight line with point A; and I also say that, with centre E having moved with the same movement to point M, point A will have come to point N and will similarly lie on a straight line with points A and G. And when point E comes to point B, point A will have traversed all the points that are along AE and will have described the line AE.

Draw a line from point H to point A. Since line FH is at right angles to line BE and so is line EA, these two lines (by Euclid, *Elements*, 1.27) will be parallel, and HA will be (by Euclid, *Elements*, 1.33) equal and parallel to FE.[51] But IG is parallel to FE; therefore (by Euclid, *Elements*, 1.30) it will be parallel and equal to AH,[52] and therefore arc AI will be equal to arc HG.[53] Therefore point A will have traversed arc HG and will have reached point G, and N and G are in a straight line for the reason given above.

We shall prove the same for point P. Since point E has come to point M, the perpendicular MPS raised from point M and cutting the quarter-circle at point P, I say that point A has come to point N. Because if point A were moved along the quarter-circle APB in the same ratio, it would have come to point P; but arc AP is equal to arc SN since the [S, 305v] line that subtends them both is the same. Therefore (by Euclid, *Elements*, 3.28 and 3.3), since the circles are equal, if one wanted to one could prove a similar conclusion from the equality of the circles and from the similitude of the movements.[54] Thus we have with two movements, one straight and the other circular, described a straight line, which was what we wanted to do. One could make instruments of all of this in which the effect could be seen very well.

AN. So as not to give you more labour, I am content for now with the demonstration given, since I have understood it very well. Would it now please you to go on to the rest?

<Proposizione V. Che i due movimenti che descrivono il cerchio sieno disgionti d' ogni proporzione in ogni tempo>

PR. Che hora i movimenti che descrivono il cerchio sieno disgionti d' ogni proportione in ogni tempo può rimanere chiaro da quel che s' è detto. Perché quando in alcuna proportione fossero per alcun tempo congionti, descriverebbono per quel tanto tempo retta et non curva. Et se di nuovo fosse detto et perché descrive la circulare curva et non l' ovale ó la parabola ó l' hiperbole, prima si può rispondere che quelle figure nella descrittione loro non sono al tutto disgiunte d' alcuna proportione; et appresso, le linee loro non sono d' una istessa lungheza. Voglio dire che quantunque la linea che descrive il cerchio si muova con due movimenti, nondimeno quel punto che descrive la circonferenza ha da moversi così conditionatamente c' ha da essere sempre con una medesima distanza del ponto del centro, il che non può avvenire nelle figure dette di sopra. Possiamo ancora giungere non essere i movimenti che descrivono le figure dette di sopra l' uno naturale et l' altro contra natura, ma tutte due s' intendono ad un modo. Ha adonque Vostra Signoria il principio mecanico d' Aristotele, detto con quel più facil modo, che sia possible.

<IV Alternativa prova del principio della meccanica>

AN. Non si poteva con altro mezo provare questo stesso?

PR. Potevasi senza dubbio, et con più facil modo; et Vostra Signoria ascolta come.

<Proposizione VI. Che il grave si muovi velocissimamente per una linea perpendiculare, e con tanto meno velocità quanto sia più lontano dalla perpendicolare>

Primieramente suppongo il movimento d' alcun grave essere, [S, 306r] mentre non è impedito, velocissimamente al[37] centro del mondo; et però per la più breve via si moverà che possible sia et questa sarà per una linea perpendiculare. Questo presupposito è manifesto dal senso, poiché può l' huomo certificaresene col senso.

AN. Questo l' ho io per chiaro et però può l' Altezza Vostra venire al resto.

37 al] all' *S*

<Proposition V. That the two movements that describe the circle are devoid of any ratio in any time>

PR. That the movements that describe the circle are devoid of any ratio in any time can now be made clear from what was said. Because if they were conjoined in any ratio through any time, they would describe in this time a straight and not a curved line. And if again it were asked why they would describe a circular curve and not an ellipse or a parabola or a hyperbola, one could reply first that these figures in their description are not entirely devoid of any ratio; and second, that their lines are not of the same length. I mean that although the line that describes a circle is moved with two movements, nevertheless the point that describes the circumference has to move in such a way that it is always the same distance from the point at the centre, which cannot happen in the other figures. We can also add that the movements that describe the other figures are not on the one hand natural and on the other against nature, but both are understood in one way.[55] You have, therefore, Aristotle's mechanical principle, expressed in the easiest way possible.

<IV Alternative Proof of the Principle of Mechanics>

AN. Can the same thing not be proved by other means?

PR. Without doubt it can, and in a very easy way; and you shall hear how.

<Proposition VI. That a heavy body will be moved most swiftly along the perpendicular, and with so much less speed the more it is removed from the perpendicular>

First, suppose that the movement of some heavy body is, [S, 306r] when not impeded, swiftest towards the centre of the world; and thus it will be moved along the shortest path that is possible, which would be along a vertical line. This supposition is obvious from sense, since a man can verify it with his senses.

AN. This is clear to me, and so you can go on to the rest.

PR. Adonque se 'l velocissimo movimento del grave sarà per la perpendicolare, rimovendosi da quella non sarà velocissimo. Così adonque intendendo noi sul piano BG la linea AB essere à piombo, et dal ponto A lassando il grave A, verrà per.quella con movimento velocissimo [*vide figuram VII*]. Ma se fosse obligato à descendere per la linea AC, verrà con meno velocità, et con meno per AD, et con meno per AK, et sempre si ritarderebbe, tanto che quando intendessimo un piano parallello al piano BG, se non havesse inclinatione ad alcuna parte, il grave per quello non si moverebbe ponto. Da qui adonque avverra che pigliando spacii equali su le linee AB, AK, et AG, et nelle altre ancora, che 'l grave più tempo consumerà à passare quello che fosse più dalla perpendicolare lontano, et però più tempo consumarebbe in passare AO che AK, et più in passare AL che AK.[38] Et se in N, K, I, et O fossero resistenti d' ugual forza meno botta farebbe nell' O che nel K, et meno nel K che nell' I. Et gagliardissima farebbe la percossa in N.

AN. Tutto questo ho inteso benissimo, et tutto è vero.

<Proposizione VII. Che il grave si muovi con maggior velocità per l' arco del cerchio maggiore che per quello del minore>

38 AL che AK.] AK che AL. *S*

PR. Therefore, if the swiftest movement of the heavy body is along the vertical, removed from the vertical it will not be swiftest. So then let us suppose that line AB is plumb to plane BG, and being released from point A, the heavy body A will fall to it with the swiftest movement (see figure VII). But if it were constrained to descend along the

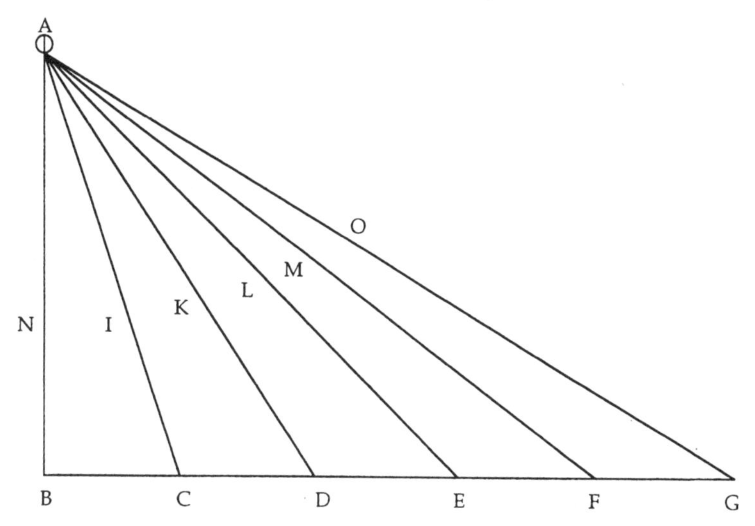

Figure VII

line AC, it would go with less speed, and along AD with even less, and along AK with still less, and ever more slowly, until, when we suppose a plane parallel to plane BG, if it has no inclination in any direction, the heavy body will not be moved along it at all. From this it follows, by choosing equal lengths on the lines AB, AK, AG and so on, that the heavy body will consume more time to traverse the line that is more removed from the perpendicular, and so it would consume more time in traversing AO than AK, and more in traversing AL than AK. And if in N, K, I, and O there were resistances of equal force, it would make less impact on O than on K, and less on K than on I. And the impact would be strongest on N.[56]

AN. All this I have understood well, and it is all true.

<Proposition VII. That a heavy body will descend with greater speed along the arc of a larger circle than along that of a smaller>

PR. Appresso Vostra Signoria ha da sapere che se 'l grave sarà obligato à moversi per cerchio, si come sono le cose delle bilanze, che per il cerchio maggiore si muoverà con maggior velocità, et questo è simile [S, 306v] al principio d' Aristotele. Però adonque intendendo noi il grave nel ponto A, come habbiamo detto, velocissimo sarebbe il movimento suo per la linea AB [*vide figuram VIII*]. Ma intendendo che s' habbia da movere obligato et in giro, con più velocità si moverà per l' arco AE che per l' arco AD, et per questo con più che per l' arco AC, intendendo però che gli spacii che haverà da passare il grave super gli archi sieno equali.

AN. Et quale è la cagione di questo?

PR. La cagione è perché il cerchio ò circonferenza del maggior cerchio meno s' allontana dalla linea retta che la circonferenza del minore – cosa che fece dire à quel valent' huomo, che 'l retto et il curvo coincidevano si come il massimo et il minimo: in infinita distanza.

<Proposizione VIII. Che la circonferenza maggiore meno s' allontana dalla linea retta che la minore>

AN. Et come proverà l' Altezza Vostra che la circonferenza maggiore meno s' allontana dalla linea retta che la minore? Perché che non possano l' una passare nell' altra lo so pur troppo, perché se 'l curvo dovesse passare in retto, si priverebbe della forma del curvo; et così il

PR. Next you should know that if the heavy body is constrained to be moved along a circle, as are things on a balance, they are moved along the greater circle with greater speed, and this is similar [S, 306v] to Aristotle's principle. Therefore, if we suppose the heavy body to be at point A, its movement, as we have said, will be swiftest along line AB (see figure VIII). But supposing it were also constrained to be moved in

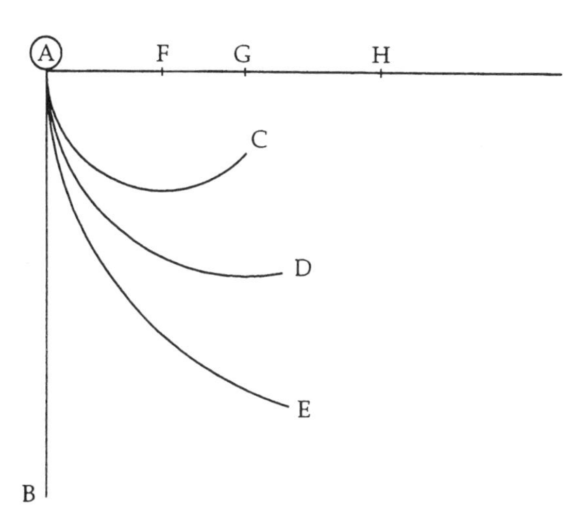

Figure VIII

a circle, it would be moved with more speed along arc AE than along arc AD, and along arc AD with more than along arc AC, assuming that the distances the heavy body would have to traverse on the arcs were equal.[57]

AN. And what is the reason for this?

PR. The reason is that the circle or circumference of the larger circle is less removed from a straight line than the circumference of the smaller – a saying attributed to that capable man, that the straight and the curved meet just as the maximum and the minimum: at an infinite distance.[58]

<Proposition VIII. That the larger circumference is less removed from a straight line than the smaller>

AN. And how will you prove that the larger circumference is less removed from a straight line than the smaller? For I know too well that the one cannot pass into the other, because if a curve could pass into a straight line, it would lose the form of a curve; and similarly a straight

retto dovendo passare nel curvo non sarebbe più retto, cosa fuggita dalle specie.

PR. Così è. Che poi la circonferenza maggiore meno s' allontana dalla retta che la minore ne darò a Vostra Signoria molte ragioni.

La prima, che se si piglieranno linee equali et si metteranno dentro à cerchi inequali, maggior circonferenza abbraccierà quella del minore che del maggiore et però maggior curvità ritenerà in se. Appresso, se si piglieranno in cerchi inequali circonferenze equali, la corda dell' arco del cerchio minore sarà minore che quella del cerchio maggiore, et per ciò ancora maggior curvità sarà nell' arco minore che nel maggiore.

Ancora, posso ció cavare dall' angolo della contingenza così. Quella circonferenza [S, 307r] è meno lontana dalla retta che fa minor angolo di contigenza, si come nell' essempio proposto poco fa può vedersi, dove minore é l'angolo di contingenza EAB che DAB et questo minore che CAB [*vide figuram VIII*]. Et però, presi archi equali nelle circonferenze dette, il grave con maggior velocità si moverà per AE che per AD, et per questo con maggiore che per AC. Giugniamo à questo, che meno è lontano l' arco AE dalla perpendicolare AB che AD, et questo meno che AC. Adonque seguirà che sempre col far maggior il cerchio potremo venir velocitando il movimento del grave.

Or per provare che due linee equali messe in cerchi inequali, che maggior circonferenza si taglierà del minore che del maggiore, farò così. Sieno due cerchi inequali et il maggiore sia A et il minore B, et sieno in essi messe due linee equali, cioè CD nel maggiore et EF nel minore [*vide figuram IX*]. Constituiscasi poi sopra la detta linea CD (per la 22ª del primo d' Euclide) un triangolo[39] isoscele, i due lati del quale sieno equali

39 un triangolo] con triangoli *S*

line able to pass into a curve would no longer be straight, having fled from its species.

PR. Like this. I shall give you many reasons why the greater circumference is less removed from a straight line than the smaller.

The first is, that if one takes equal lines and puts them inside unequal circles <i.e., as chords>, the arc of the smaller <circle> will consist of a larger <part of the> circumference than that of the larger and will therefore contain a greater curvature. Again, if one takes equal arcs in unequal circles, the chord of the smaller circle will be less than that of the larger circle, and for this reason again greater curvature will be in the smaller arc than in the larger.

Again, I can educe this from the angle of contingence thus. The arc [S, 307r] that makes a smaller angle of contingence is less removed from a straight line, as one can see in the example just above, where the angle of contingence EAB is less than DAB and DAB less than CAB (see figure VIII).[59] And therefore, taking these arcs equal in length, the heavy body will be moved with greater speed along AE than along AD, and along AD with greater speed than along AC. Let us add to this that the arc AE is less far from the perpendicular AB than AD, and AD is less far than AC. Thus it follows that by making the circle ever larger we can speed up the movement of the heavy body.

Now, to prove that when two equal chords are put into unequal circles, a larger arc will be subtended in the smaller than in the larger, I shall proceed thus. Let there be two unequal circles and let the larger be A and the smaller B, and let there be put into them two equal chords, CD into the larger and EF into the smaller (see figure IX). Then erect on CD (by

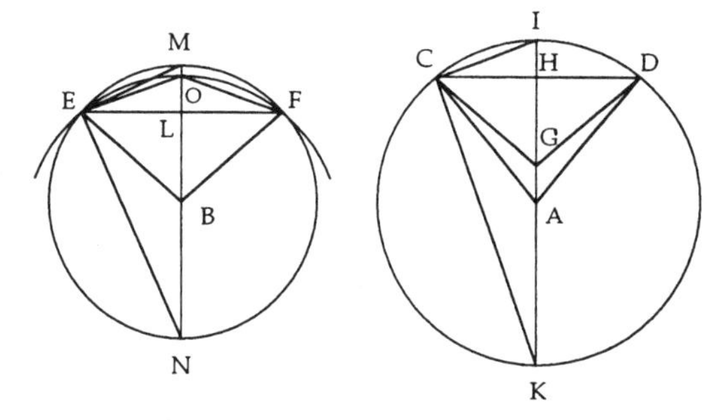

Figure IX

alle linee BE et BF. Si può tal triangolo fabricare poiché la linea EB è minore della linea AC[40] et sia quello CGD. Habbiamo adonque (per la 21ª del primo d' Euclide) l' angolo CGD essere maggiore dell' angolo[41] CAD. Or produchisi la linea perpendicolare dal ponto A sopra la linea CD et distendasi fino alla circonferenza dall' una et l' altra parte, et sia IAK. Produchisi ancora su la linea EF <la linea perpendicolare> dal <ponto[42]> B et sia MBN. Non è dubbio, così la linea IAK taglierà et l' arco CID et l' angolo CGD et la linea CHD, come la linea MBN tagliare et l' arco EMF et l'angolo EBF et la linea [S, 307v] ELF in due parte equali; et però così l' angolo ELB come l' angolo CHA essere (per la terza del terzo d' Euclide) retto. Producasi dal ponto C al ponto I la retta CI et dal ponto C al ponto K la retta CK; così dal ponto E al ponto M la linea EM et dal ponto[43] E al ponto N la linea EN. Sarà adonque (per la 31ª del terzo d' Euclide), così l' angolo ICK come l' angolo MEN retto. Habbiamo hora nel triangolo ENB il lato NB prodotto; adonque l' angolo EBM sarà equale (per la 32ª del primo) gl' angoli BNE et BEN. Ma i detti angoli sono (per la quinta del primo) equali; adonque l' angolo EBM è doppio all' angolo BNE. Per le medesime ragioni l' angolo CAI è doppio all' angolo AKC. Ma l' angolo EBM è maggiore dell' angolo CAI, essendo il tutto maggiore, et l' uno et l' altro è la mita del suo tutto; adonque l' angolo BNE è maggiore dell' angolo AKC. Ma gli angoli MNE et NME sono equali ad un retto, et così gli angoli IKC et KIC; adonque l' angolo KIC sarà maggiore dell' angolo NME.

Di nuovo, gli angoli MLE et IHC sono retti; adonque, così gli angoli LME et LEM sono equali ad un retto, come gli angoli HIC et HCI. Ma è stato provato l' angolo KIC maggiore dell' angolo NME; adonque resterà l' angolo LEM essere maggiore dell' angolo HCI. Or sopra il ponto E della linea LE (per la 23ª del primo d' Euclide) descrivasi l' angolo LEO equale all' angolo HCI et congiongasi OF, la quale (per l' ottava del primo) sarà equale all' OE. Sopra 'l dato triangolo EOF ò intorno à quello (per la 5ª del quarto d 'Euclide) descrivasi la portione del cerchio EOF. Tal portione di necessità sarà equale alla portione CID, poiché sono sopra linee equali et hanno gli angoli CID et EOF equali (per la 24ª del terzo d' Euclide). Che le linee sieno equale è manifesto del presupposito; et che l' angolo CID sia equale all' angolo EOF è ancor manifesto per la 32ª d' Euclide. Contiene

40 AC] EC *S*
41 angolo] ango *S*
42 ponto *lacuna in S*
43 ponto] pon *S*

Euclid, *Elements*, 1.22) an isosceles triangles, the two sides of which are equal to lines BE and BF.[60] One can make such a triangle since line EB is less than line CA: let it be CGD. We have it then (by Euclid, *Elements*, 1.21) that angle CGD is greater than angle CAD.[61] Now, draw a line from point A perpendicular to line CD and produce it to the circumference at both ends, and let it be IAK. Again, draw <a line> from point B <perpendicular> to line EF and let it be MBN. There is no doubt that, just as line IAK will cut arc CID, angle CGD, and line CHD into two equal parts, so line MBN will cut arc EMF, angle EBF, and line [S, 307v] ELF into two equal parts; therefore angle ELB, as angle CHA, (by Euclid, *Elements*, 3.3) will be a right angle.[62] From point C to point I draw the straight line CI and from point C to point K the straight line CK; similarly from point E to point M draw the line EM and from point E to point N the line EN. Therefore, (by Euclid *Elements*, 3.31) angles ICK and MEN will be right.[63] We have now made in triangle ENB the side NB; therefore angle EBM will be (by Euclid, *Elements*, 1.32) equal to <the sum of> angles BNE and BEN.[64] But these angles are (by Euclid, *Elements*, 1.4) equal;[65] therefore angle EBM is twice angle BNE. For the same reasons angle CAI is twice angle AKC. But angle EBM is larger than angle CAI, since the whole is larger and each <i.e., angle BNE and angle AKC> is the half of this whole; therefore angle BNE is larger than angle AKC. But angles MNE and NME are <in sum> equal to one right angle, and so also are angles IKC and KIC; therefore angle KIC will be larger than angle NME.

To begin anew, angles MLE and IHC are right; therefore, just as angles LME and LEM are <in sum> equal to a right angle, so are angles HIC and HCI. But angle KIC was proven larger than angle NME, so it follows that angle LEM is larger than angle HCI. Now, at point E of line LE (by Euclid, *Elements*, 1.23), draw the angle LEO equal to angle HCI and join OF, which (by Euclid, *Elements*, 1.8) will be equal to OE.[66] On the resulting triangle EOF or around it (by Euclid, *Elements*, 4.5) draw the circular arc EOF.[67] This arc of necessity will be equal to the arc CID since they are on equal chords and have angles CID and EOF equal (by Euclid, *Elements*, 3.24).[68] That the chords <EF and CD> are equal was presupposed; that the angle CID is equal to angle EOF is again clear by Euclid, *Elements*, 1.32.[69] Now arc EMF contains arc EOF; but the container is larger than the contained; therefore arc EMF [S, 308r] will be

hora la portione EMF la <portione[44]> EOF; ma il continente è maggiore del contenuto; adonque maggiore sarà la [S, 308r] circonferenza EMF della circonferenza CID. Ma sono sopra una stessa corda; adonque più curva è la portione EMF che la portione CID, ch' è quello che volevamo.

AN. Bella dimostratione è stata questa, ma non poteva l' Altezza Vostra dimostrare l' istesse con altro mezo?

PR. Poteva si, ma con più difficoltà; oltre che, sarebbe stata di mia inventione come questa. Con questa dimostratione si può ancora dimostrare quello c' ha detto Aristotele et che noi poco fa habbiamo dimostrato del resto ò residuo del diametro.

<Proposizione IX. Che di eguale archi de disuguali cerchi, quella nel cerchio maggiore sottotenderà maggiora corda, che quella nel minore>

Da questa ancora come per corollario possiamo cavare che, se noi piglieremo equale circonferenze nel maggiore et nel minore cerchio, che alla circonferenza del maggiore si sottotenderà maggior linea retta, che sotto la circonferenza del minore. Come sia il minor cerchio ABEC [*vide figuram X*] sopra il centro F, il maggiore sia CGK sopra il centro D, et sia la parte della circonferenza CAB del minore sopra la linea retta BC, sopra à

44 portione *lacuna in S*

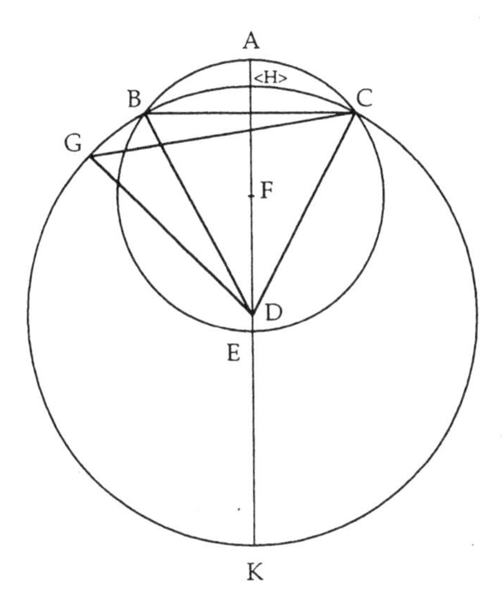

Figure X

larger than arc CID. But they are on the same chord; therefore arc EMF is more curved than arc CID, which is what was desired.[70]

AN. This was an elegant demonstration, but could you not demonstrate the same thing by another means?

PR. I could, but with more difficulty; besides which, it would be of my own invention, as this was. With this demonstration one can also demonstrate what Aristotle said and what we have just demonstrated of the rest or residual of the diameter.[71]

<Proposition IX. That of equal arcs in unequal circles, the one in the larger circle subtends a longer chord than the one in the smaller>

From this conclusion again as a corollary we can educe that, if we take equal arcs of a larger and of a smaller circle, the arc of the larger will subtend a longer straight line than the arc of the smaller. Let the smaller circle ABEC be around centre F, let the larger be CGK around centre D, and let arc CAB of the smaller be on the straight line BC, on which also let arc CHB of the larger be (see figure X). It is clear from the

quale ancora sia la parte CHB della circonferenza del maggiore. È chiara cosa per la precedente dimostratione la circonferenza CHB essere minore della circonferenza CAB. Presupponghiamo la circonferenza CBG essere equale alla circonferenza CAB, et tirisi dal ponto C al ponto G la linea CG essere maggiore della linea CB. Tirisi dal centro D le linee DC, DB, et DG. Perché la circonferenza CHB[45] è minore della circonferenza CG, adonque (per la 26ª[46] del terzo d' Euclide) l' angolo CDB nel centro sarà minore dell' angolo CDG, et (per la 24ª del primo dello stesso autore) la base CB sarà minore ne triangoli [S, 308v] DCB et DCG della base CG, ch' è quello che si voleva dimostrare.

<V Conclusione>

AN. Ho inteso benissimo il tutto, ma che cosa si vuole da questa inferire?

PR. Dirò à Vostra Signoria tutte le precedenti demostrationi sono state fatte ad effetto per dimostrare che 'l minor cerchio, con la maggior curvità, ma più contraviene al movimento del grave, che 'l maggiore, non considerando ponto la lungheza del viaggio. Perché già vede Vostra Signoria che dovendosi movere un grave per l' arco CAB[47] [*vide figuram* X] con più fatica si moverà per la curvità sua che per l' arco CHG à quello equale, et per conseguente se le leva molto della forza et del vigor suo. Et levasele ancora del peso, poiché si può dalla velocità del movimento aumentare[48] il peso. Ma di questo appresso. Resta adunque per manifesto dalle dimostrationi precedenti, che quanto meno[49] un peso obligato à moversi in giro, ò una forza s' allontana dal centro, con tanto maggior velocità si moverà et la forza tanto maggior effetto farà.

AN. Ho capito il tutto et resto del tutto capace. Ma favoriscami Vostra Altezza a dirme se tutte le cose che si fanno con forza meccanica dependono da questo solo principio?

PR. Buona parte depende da questo solo, ma molte altre dependono dalla forza del vacuo, come sono molte machine d' acqua; molte dall' aria moltiplicare, come sono le mine et l' arteglierie; molte dalla ragione delle figure, come sono le cose da tagliare, et delle ordinanze, et insieme delle forteze et delle navi.

45 CHB] CB *S*
46 26ª] 27ª *S*
47 CAB] ACB *S*
48 aumentare] argomentare S
49 meno] più *S*

preceding demonstration that arc CHB is less than arc CAB. Let us suppose that arc CBG is equal to arc CAB, and draw from point C to point G the line CG larger than line CB.[72] Draw from centre D the lines DC, DB, and DG. Because arc CHB is less than arc CG, (by Euclid, *Elements*, 3.26) angle CDB in the centre will be less than angle CDG;[73] and (by Euclid, *Elements*, 1.24) the base CB of triangle DCB will be less [S, 308v] than the base CG of triangle DCG, which is what was to be demonstrated.[74]

<V Conclusion>

AN. I have understood it all very well, but what can be concluded from it?

PR. All the preceding demonstrations were made for the purpose of demonstrating that the smaller circle, with its greater curvature, more contravenes the movement of the heavy body than the larger circle, without considering at all the length of the motion. For you have already seen that a heavy body, constrained to be moved along arc CAB [see figure X], is moved with more effort along its curvature than along arc CHG equal to it, and consequently it loses much of its force and vigour. And it also loses weight, because weight can be augmented by speed of movement. But of this later.[75] Thus it is clear from the preceding demonstrations that the less a weight is constrained to move in a circle, or the farther a force is from the centre, with so much more speed it will move and so much more effect the force will have.

AN. I have understood it all and I rest entirely content. But would you favour me by saying whether every effect that is made with mechanical force depends on this principle alone?

PR. A good part depends on this alone, but many others depend on the force of the vacuum, such as many water machines. Many <depend> on the multiplication of air, such as mines and artillery; many on the design of their shapes, such as things that cut, ordnance, and both fortification and ships.

AN. Adonque la mecanica si distende nelle cose della militia. Un mio che mi leggeva Euclide con autorità di Proclo, mi diceva la militia non si comprendere è sotto le matematiche, et però non sotto la meccanica.

PR. È vero che Proclo tiene di no, ma vi sono ragioni potentissime per il contrario. Poiché noi sapiamo il fine del meccanico essere di superare le molte forze con le poche, ò con le poche resistere alle molte; ma il soldato fa questo per virtù della figura dell' ordinanza sua. Ma tal consideratione della figura è del meccanico, et però la militia et in questa dell' ordinanze et in molte altre è sotto[S, 309r]posta alla mecanica.

AN. À quel ch' io veggo sono molte le parti della mecanica.

PR. Molte sono per certo. La prima è quella delle machine da ruote; la seconda quella delle machine d' acqua; la terza l' arte di far le navi; la quarta l' arte delle fortezze; la quinta l' arte[50] da far l' arteglierie, delle mine, et de fuochi artificiali; la sesta la parte della militia che tratta dell' ordinanze. Sonovi ancora molte altre parti d' essa che non mi sovengono: ma in somma dove s' adopera alcuna machina da far forze, tutta tal parte è mecanica et vien compresa d' alcuna delle dette parti. La scienza di pesi è sotto la mecanica, scienza di molto momento, della quale un' altro giorno à Dio piacendo ragioneremo. Nella detta scienza si comprende la cosa d' Archimede delle cose che pesano equalmente.

AN. Io vorrei ancora domandare alcun' altra cosa all' Altezza Vostra, ma parmi che sia hora ch' ella vadi à pigliar aria, et però mi porrò silentio à medesimo.

PR. Vostra Signoria può ancora dire alcun' altra cosa, et poi se le parerà essendo com' è tardi, andaremo à cavalcare per la città.

AN. À tutte le cose dette basta solo il principio detto, ò pure vi sono altri principii?

PR. Il principio dechiarato fin qui è universale à quasi tutte quelle cose che si possono ridurre alla stanga ó lieva ó vecte, ma l' altre hanno i loro principii particolari. Et molte ancora che si servono della stanga hanno i loro principii proprii.

AN. Il principio detto fin qui può condurci alla cognitione d' alcun problema?

PR. Anzi di molti.

AN. È quando sarà servita l' Altezza Vostra di favorirme à dirmene alcuno?

50 l' arte] l' Artre *S*

AN. Mechanics then extends into military concerns. A <retainer> of mine who read me Euclid with Proclus's *Commentary* told me that the military <arts> are not to be understood as under mathematics and thus not under mechanics.[76]

PR. It is true that Proclus denies it, but there are most powerful reasons to the contrary. For we should note that the goal of the mechanic is to overcome great forces with small, or with small forces to resist great; but the soldier does this by virtue of the deployment of his forces. Such a consideration of design is proper to the mechanic, and therefore the military both in the matter of deployment and in many other respects is subordinate [S, 309r] to mechanics.[77]

AN. From what I have seen the parts of mechanics are many.

PR. There are certainly many. The first concerns wheeled <i.e., geared> machines; the second water machines; the third the art of making ships; the fourth the art of fortification; the fifth the art of making artillery, mines, and artificial fire; the sixth the part of the military arts that treats of deployment. There are also many other parts of it that do not occur to me: but in sum, wherever some machine is applied to make force, every such part is mechanical and is included in one of those parts above.[78] The science of weights is under mechanics, a science of much importance, which, if it pleases God, we shall discuss another day. In that science is included the work of Archimedes *On Plane Equilibrium*.[79]

AN. I should still like to ask you something more, but it seems to me that it is the hour that you go to take the air, and therefore I shall remain silent.

PR. You can still say something more, and then if you think it is too late, we shall go ride through the city.

AN. Does the single principle we stated suffice for all the things that have been mentioned, or are there still other principles?

PR. The principle expounded thus far is universal to almost all those things that can be reduced to the lever, but the others have their own particular principles. And even many that use the lever have their own proper principles.

AN. Can this principle guide one to the solution of any problem at all?

PR. Of most, in fact.

AN. And could you favour me by telling me some of them?

PR. Quando à Vostra Signoria parerà.

AN. Questo non è favore da riceverlo con lungheza di tempo, ma con presteza: et però se Vostra Altezza è servita, io me ne verrò da buon discepolo dimane all' hora d' hoggi.

PR. Vostra Signoria sarà la ben venuta, et la sua venuta mi sarà di molto contento, poiché verrò ricordandomi molti problemi mecanichi [S, 309v] che m' havevo dimenticato, et insieme verrò rivedendo i dissegni et i modeli di molte machine c' ho in un' altro luogo per poterle mostrare à Vostra Signoria, de quali so che riceverà molto contento perché vedrà in quelli la forza delle cose mecaniche et à quanto s' è disteso l' ingegno dell' huomo.

PR. If you like.

AN. It is no favour to receive in the fullness of time, but rather promptly; and therefore, if it suits you, I shall be a good pupil tomorrow at the same time as today.

PR. You will be welcome to it, and your coming will be very satisfying to me, since I shall have remembered many mechanical problems that [S, 309v] I have forgotten, and I shall also look over the designs and models of many machines that I have in another place in order to be able to show them to you, from which I know you will receive much satisfaction because you will see in them the power of mechanical things and how far the genius of man extends.

<GIORNATA SECONDA>
<I Principii del Moto>

AN. Hiersera doppo ch' io lasciai l' Altezza Vostra, non mi parendo haver fatto esercitio a mio modo, mi messi a caminare fuori Te, et viddi alcune canne in fasci, le quali forse erano quivi per farne ó alcun siepe ó per coprirne alcuna capana. Intorno a queste canne erano alcuni fanciulli che le lanciavano à gara ad una ad una, ma con una destreza mirabile. Ciascun d' essi haveva un' altro pezo di canna in mano, et in cima di quella era un' intalio ò foramine[1] accommodato con arte, nel quale appoggiavano il piede della canna che lanciar volevano, et con l' altra mano tenevano la canna nel mezo, et così le lanciavano lontanissimo. Ma pareva che chi voleva più sfozarsi le lanciava meno. Et se per ventura in alcuna verso la cima fosserò state foglie le[2] toglievano, anzi tagliavano alquanto della punta. Volli vedere se l' istesso havessero potuto far gl' huomini, et così da due miei staffieri fece fare l' istesso; ma in fatti, subbito che le canne uscivano dalle lor mani, tremando cadevano poco lontane. La dove io son venuto pensando, che tal cosa fosse pertinente alla mecanica et però ho voluta dirla all' Altezza Vostra.

PR. È per certo cosa mecanica, et Aristotele nelle sue *Questioni <meccaniche>* ne rende la cagione, et io appresso la dirò a Vostra Signoria.

AN.[3] Bene adonque ch' io aspetti ad udirla a suo loco, poiché presto vi verremo.

PR. Non lasciaremo [S, 310r] di ragionare hoggi c' havremo dato risposta, così a questa come a molte dimande. Crederò che Vostra Signoria si ricorda il principio che hieri fu dichiarato da noi et stabilito con le sue dimostrationi?

1 foramine] forame *S*
2 le] Le le *S*
3 AN.] AD. *S*

<THE SECOND DAY>
<The Principles of Motion>

AN. Last evening after I left you, not having had my usual exercise, I went for a walk outside the Palazzo Te,[1] and I saw some reeds in bundles that perhaps were there to make hedges or to cover some huts. Around these reeds were some boys who were throwing them in competition with one another, but with marvellous skill. Each of them had another piece of reed in his hand, at the top of which there was a carefully made notch or hole into which they placed the tip of the reed that they wanted to throw, and with the other hand they held the reed in the middle, and in this way they were throwing them very far. But it seemed that those who applied more force threw less far. And if by chance there were leaves near the end, they removed them – in fact they cut off some of the point. I wanted to see if men could do the same, and so I had two of my footmen try; but in fact, as soon as the reeds left their hands they vibrated and fell a short distance away. Whence I got to thinking that such a phenomenon was pertinent to mechanics and so I wanted to tell you about it.

PR. It certainly pertains to mechanics, and Aristotle in his *Mechanical Problems* gives the cause, and I shall describe it to you later.[2]

AN. So it would be well for me to wait to hear it in its proper place, since we shall soon come to it.[3]

PR. We shall not omit [S, 310r] to discuss today our answer to this as well as to many other questions. I trust you remember the principle that we stated yesterday and established with its demonstrations?

AN. Me lo ricordo benissimo; anzi son' venuto questa notte et hoggi epilogandolo a me stesso. Et però Vostra Altezza può venire al resto.

PR. Perché meglio possiamo considerare le cose che concorrono alle forze mecaniche, proponemo un problema et intorno à quello verremo considerando tutte ò la maggior parte delle minutie che concorrono a far bene una operatione mecanica. Diremo adonque: Ond' è che un gran peso, che non si può con le mani et senza stanga ò lieva da molti huomini muovere, con la stanga vien da meno numero mosso, essendo che al peso si giugne peso il quale è quello della stanga?

AN. Io saprei rispondervi col principio dichiarato hieri et direi che l' essere la forza più lontana dal centro viene a descrivere maggior cerchio, et però con più velocità si muove, et però più pesa.

<I L' immobile>

PR. Vostra Signoria ha dato quasi al segno, ma bisogno intorno a tal domanda giugnere alcune cose, le quali sono queste. Primieramente Vostra Signoria vede due cose principali: il peso da moversi et la forza che l' ha da movere. Oltre a queste[4] vede la stanga, senza la quale dal presupposto non havrebbe potuto farsi il movimento del peso. Ma oltre queste tre cose v' è il luogo dove s' appoggia la stanga; perché se non si fermasse in alcun luogo non si potrebbe far la forza che si vuole, ó perché ogni cosa c' ha da violentare un' altra è forza che s' appoggi in alcun luogo fermo ó ch' essa sia ferma. Et però disse Archimede che s' havesse havuto luogo dove havesse potuto appoggiar la machina ch' havrebbe tirato questo mondo in quello. Il che ha inteso Aristotele nel libretto del *Movimento de gl' animali,* dove dimostra, che nell' animale stesso sono le gionture per centro del movimento delle membra. Percioche prima dimostra che ciascuna cosa che si muova ha da moversi sopra [S, 310v] alcun immobile, il quale immobile ó è dentro la cosa stessa ó è fuori. Ne gli animati, mentre che voglion muovere le membra, il principio stabile è in loro stessi; percioche mentre che s' ha da muovere il braccio nell' huomo sta ferma la spalla, et mentre che si muove la mano è ferma una parte del braccio. Nel muoversi poi, che fa l' huomo da luogo a luogo, è fermo il solo dove s' ha da movere, come la terra. Et però mentre che 'l solo sopra il quale s' ha da muovere l' huomo non è saldo, il movimento non si può fare con quella velocità che bisognerebbe. Da dove è che meno si camina super la sabbia, per la neve, et

4 queste] questo *S*

AN. I remember it very well; in fact I have come tonight rehearsing it to myself. Therefore you can proceed to the rest.

PR. So that we might better consider the things that contribute to mechanical forces, I shall propose a problem and we shall proceed to consider all or most of its details that contribute to producing a mechanical effect. We shall ask, then: How is it that a large weight, which cannot be moved by the hands of many men without a lever, can be moved by fewer with a lever, even though the weight of the lever is added to the load?[4]

AN. I should reply with the principle stated yesterday and say that, since the force is farther from the centre, it comes to describe a larger circle, and thus it is moved with more speed and so exerts more weight.

<I The Immobile>

PR. You have almost hit the mark, but I need to add to this question several things, which are as follows. First, you see two principal things: the weight to be moved and the force that must move it. Besides these you see the lever, without which (by supposition) the weight could not be moved. But besides these three things there is the place where the lever rests; because if it were not fixed in some place one could not exert the force one wanted, or because everything that is to coerce something else is a force that rests on some fixed place or is itself fixed. And therefore Archimedes said that if he had a place where he could put the machine he could draw this world to it.[5] With which Aristotle concurred in his book on *De motu animalium*, where he shows that in the animal itself there are joints providing the centre of movement of the members.[6] Therefore he first shows that everything that is moved must be moved against [S, 310v] some immobile that is either inside the thing itself or is outside. In animate things, when they want to move their members, the stable principle is within themselves; for when a man's arm is moved the shoulder remains fixed, and when the hand is moved a part of the arm is fixed. In moving himself, then, which a man does from place to place, the ground where he moves is fixed, as the earth. And therefore when the ground on which the man must move is not solid, the movement cannot be made with the speed that might be desired. Whence it is that one walks less <quickly> on sand, through snow, through mud, and

per il fango, et sopra il terreno ó dall' arratro ó dalla zappa ó d' altro mosso, che sopra il fermo. Concorrono adunque nel movimento locale de gl' huomini et de gli animali questi due principii stabili, cioè quello che intrinseco vien detto et particolare, et l' estrinseco et universale, senza i quali non si potrebbe fare il movimento.

A. Queste son cose che apparono al senso et sono manifeste a tutti che vogliono considerarvi su alquanto. Ma le biscie, che non hanno membra, come si muovono et tuttavia venno in atto, et alcune ne sono che ascendono gli alberi?

P. Le biscie hanno ancora i loro principii stabili intrinsechi oltre poi la terra, percioche mentre che si piegano, vengono con la piegatura à far centro et immobile sopra il quale muovono il resto. Come se intenderemo il serpe ABCDE [*vide figuram XI*], che s' habbia da movere egli, come s' osserva. Si piega et fa gli angoli quasi et nel far gli angoli vieno a far i suoi luoghi fermi, come ABC, CDE, dove la parte AB si ferma nel ponto B et la parte BC nel ponto C, et così del resto. Ma come poi di queste parti l' una si muti di destra in sinistra et di sinistra in destra Aristotele lo mostra nel libro dell' *Incesso de gl' animali*, et noi appresso, per esser in cose pertinenti al mecanico, ne faremo [S, 311r] alcune consideratione. Quanto poi al dire che i serpenti si son veduti ascendere gl' alberi, io non nieghero il fatto, havendo osservato l' ascenso loro su luoghi eminenti per sassi quasi nudi. Dirò poiché tale ascenso si fa prima con le piegature dette di sopra; et oltre, il serpente ha sotto al ventre per tutto le squame non altrimente come sono le lame, che si mettono nelle coraze che l' una è sopra posta all' altra, et mentre che ascendono si servono di quelle squame in luogo di unghie attaccandole ó all' albero ó al sasso per non scivolare[5] giù da quello.

A. Bella consideratione è questa delle squame; ma il vento che pinge le navi inanti con tanta velocità, a chi s' appoggia?

P. Appunto Aristotele intorno a questo vi dice alcune parole ma non

5 scivolare] isdricciolare *S*

on ground disturbed either by the plough or by the hoe or by something else, than over solid ground.[7] These two stable principles contribute to the local movement of men and animals, viz. the one that comes to be called intrinsic and particular, and the other extrinsic and universal, without which movement could not be made.

AN. These are things that appear to the senses and are obvious to all who wish to consider the matter a little. But snakes, which do not have limbs, how do they move and come into action at all – and there are some of them that climb trees?

PR. Snakes again have an intrinsic stable principle besides the earth, because when they bend, they make with the bend a centre and an immobile against which the rest moves. As <for example> if we imagine the snake ABCDE (see figure XI), which moves itself. As you see, it

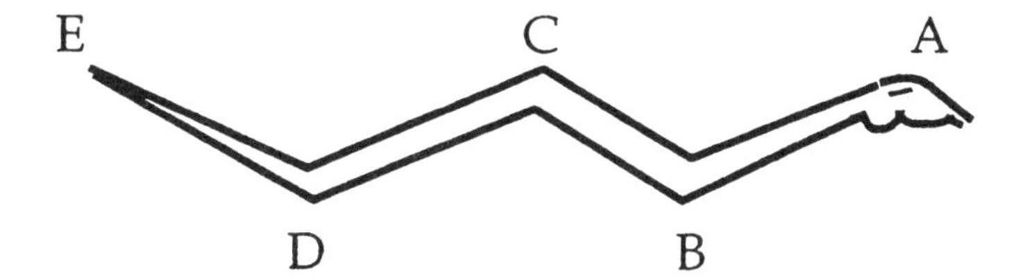

Figure XI

bends and almost makes angles and in making angles comes to make its fixed places, such as <the angles> ABC and CDE, where part AB is fixed at point B and part BC at point C, and so on. But how one of these parts moves from right to left and from left to right Aristotle showed in his book *De incessu animalium*, and since we are discussing things pertaining to the mechanic, we shall make [S, 311r] some remarks about it later.[8] As for saying that serpents have been seen climbing trees, I do not deny the fact, having observed their ascent to high places up almost bare rock. I would say then that such climbing is done first with the bending described above; and further, snakes have scales like blades covering their undersides, which are arranged in a cuirass such that one overlaps the other, and when they climb they use these scales in place of claws by attaching them to the tree or stone so as not to slip down from it.

AN. This account of scales is elegant; but the wind that pushes boats forward with such speed, to what is it fixed?

PR. On precisely this Aristotle says a few words but not many, raising it only when he discusses the extrinsic stable principle, saying that a

molte, et piglia pur occasione dal principio stabile estrinseco, dicendo che l' huomo ch' ha da pingere una nave dalla riva appoggia un remo ò tale stromento, essendo egli su la riva, all' albero ò ad altra parte della nave, et così essendo immobile la riva, è forza che la nave si muova inanti. La dove se quel tale ch' appoggia il remo alla nave et è alla riva l' appoggiasse al piede delo albero, essendo egli in nave s' havesse la forza di tutti i giganti del mondo per questo non moverebbe punto la nave. Così ancora, se 'l vento fosse dentro la nave non moverebbe punto la nave, quantunque soffiasse nel più gagliardo modo ch' egli potesse. È adonque il vento che pinge la nave innanzi in parte fuori di quella, et per conseguente s' appoggia nell' aria immobile che le sta dietro. Et se Vostra Signoria soggiognesse, che al movimento del vento si muove l' aria, al meno tanto d' essa quanto è la lingua del vento, io nol nieghero; ma alla fine di dietro ancora all' aria mosso è l' aria stabile, poiché non è provato ne dall' esperienza ne dalla ragione che l' aria si muova al movimento del vento circolarmente per tutto il circuita della terra. Anzi l' osserva il vento alcune [S, 311v] volte venire d' alcun determinato luogo come nella bocca de fiumi, et fuori d' alcun vallone tra due monti, et da luoghi tali oltre à quali non è vento alcuno. Et i naviganti col vedere una nuvoletta ò cosa tale diranno da quel luogo haveremo vento, et debile et gagliardo, secondo la qualità del nuvolo. S' appoggia adonque il vento nell' aria che le sta dietro, et così fa movere[6] la nave.

A. Del vento son per hora sodisfatto. Ma chi nuota, dove s' appoggia?

P. Vostra Signoria come huomo che sa nuotare può darne il dove. Chi nuota ponta i piedi nell' acqua, la dove se l' acqua non resistesse al pontar de piedi il nuotare andarebbe al fondo. Et però quell' acque che son più grosse sono ancora più atte ad esser nuotate. Et ció per due ragioni: l' una perché sostentano più il nuotatore con la grosseza loro, percioche non ciedeno così all' affondar di quello; et appresso perché è più ferma al pontar de piedi del nuotatore. Et però quei che sono valorosi nel nuotare nell' acque dolci valorosissimi riescono poi nelle salse, ma non al converso.

A. Ho inteso, et da qui mi vien la solutione di quel dubbio del uovo che messo nell' acqua dolce va al fondo, et nella salsa vi nuota: perché la salsa con la grosseza sua resiste al descendere dell' uovo. Et allo stesso modo può rispondermi Vostra Altezza intorno al volare de gli uccelli, percioche portano i piedi nell' aria et con le penne si sostentano in quella, et col muovere l' ali vanno oltre. Ma dicami Vostra Signoria,

6 *ante* movere *scrib. et del.* nave *S*

man standing on the bank who has to push a boat from shore sets an oar or similar instrument on the mast or other part of the ship, and so, since the bank is immobile, the ship is forced to move forward. Whence if the one who is on the bank and who pushes the ship with the oar should step to the foot of the mast, being now in the ship he could have the force of all the giants of the world and still not be able to move the ship at all. So again, if the wind were within the ship it would not move the ship at all, even if it blew in the most violent way it could.[9] It is thus the wind that pushes the ship forward from outside of it, and consequently it is fixed in the immobile air that stands behind. And if you should suggest that, when the wind blows, the air – at least as much of it as is the tongue of the wind – is moved, I would not deny it; but in the end behind the moved air there is still stable air, since it has been proved neither by experience nor by reason that when the wind blows the air is moved circularly around the entire earth. In fact I have observed the wind some-[S, 311v] times coming from some specific place, as at the mouth of rivers and out from a valley between two mountains, while from places beyond these there is no wind. And sailors when they see a little cloud or some such thing will say from what direction we shall have wind, whether weak or strong, according to the quality of the cloud. The wind, then, is fixed in the air that stands behind, and so makes the ship move.

AN. Concerning the wind I am satisfied for now. But where are swimmers fixed?

PR. You as a man who knows how to swim can say where. Swimmers kick their feet in the water, so that if the water did not resist the kicking the swimmers would go to the bottom.[10] And thus waters that are denser are even more fit for swimming. And this for two reasons: the one because with their density they support the swimmers more, since they do not yield so much to their sinking; and second because they are firmer to the kicking of the swimmer's feet. And therefore those who are strong swimmers in fresh water succeed most valiantly in salt, but not the converse.

AN. I understand, and from this comes to me the solution to the doubt over why eggs put in fresh water go to the bottom and float in salt water: because salt water resists with its density the descent of the egg. And likewise you can answer me concerning the flight of birds, since they carry their feet in the air and with their feathers sustain themselves

stima che nel volare de gli uccelli, et nel nuotare de pesci, sia cosa appartenente alla mecanica?

P. Anzi il mecanico, come appresso mostrero a Vostra Signoria, ha da movimenti loro compreso molte cose utilissime all' uso.

A. Adonque Vostra Altezza ne dirà alcuna cosa?

P. Dirone secondo l' occasione et a proprii luoghi.

A. Aspetterò adunque il tempo.

P. Ha adonque veduto Vostra Signoria che oltre il principio immobile intrinseco per il movimento de gli animati è stato forza esservi al principio immobile estrinseco. Or nelle cose della [S, 312r] mecanica è di molto momento questo principio immobile, et Aristotele lo chiama col nome d' *hypomochlion*[7]; altri l' hanno detto *foleimento*, che l' uno et l' altro vuol dire <*sottoleva*[8]>.

<II La virtù movente e la resistenza>

Seguiremo ancora tuttavia la consideratione delle qualità et delle proprietà delle quattro cose dette, et primieramente comincieremo dalla forza, ò dalla virtù c' ha da movere: la forza, in tanto vien detta forza, et virtù movente, in quanto che fa altrui forza et è cagione dell' altrui movimento, et in queste cose corporee non può una cosa far muovere un' altra ch' ella non si muova in alcun modo. (Ma accioche non siamo presi in parole, quando parliamo del movimento intendiamo del locale, perché d' altro non intendiamo noi per hora.) Et se Vostra Signoria mi dicesse che mentre un' huomo tira un sasso con la mano ch' egli non si muove punto, direi questo esser falso; perché et se non muove il resto della persona, il che non è però vero mutue, tuttavia parte di quella <muove>, percioche muove 'l braccio. Et però bene dicono i filosofi, che *omne agens naturale in agendo repatitur*. Adonque per dirsi veramente, forza et virtù movente ha da far altrui forza et da movere la cosa. Et però ciascun movimento viene ad haver due cose principali: la cagione movente et da dove è il principio del movimento; et il mobile, nel quale come in proprio soggetto è fondato il movimento. Or questa virtù movente, ó è intrinseca nel mobile ó è estrinseca. La virtù intrinseca è cagione di movimento naturale, si come la gravità nel movimento all' ingiù del sasso et la leggereza nel fuoco, che sono le proprietà che se-

7 hypomochlion] Hippomeclion *S*
8 sottoleva *lacuna in S; vide* Danielo Barbaro, tr., *I dieci libri dell' Architettura di M. Vitruvio*, X. cap. viii, p. 259.

in it, and by moving their wings go forward. But tell me, do you think that the flight of birds and the swimming of fish pertain to mechanics?

PR. In fact the mechanic, as I shall later show you, has from their movement grasped many most useful things.

AN. So will you mention some of them?

PR. I shall discuss them as appropriate and at the proper place.

AN. Then I shall await the time.[11]

PR. You have seen, then, that besides the intrinsic immobile principle for the movement of animate creatures, it is necessary for there to be an extrinsic immobile principle. Now, in things concerning [S, 312r] mechanics this immobile principle is of great moment, and Aristotle called it by the name of *hypomochlion*; others have called it *foleimento*, both of which mean <fulcrum>.[12]

<II The Moving Power and the Resistance>

In any case we shall proceed to consider the qualities and the properties of the four things mentioned before, and let us begin first with the force or power that moves <something>: force, in so far as it comes to be called force, and power of moving in so far as it exerts force on another and is the cause of its movement. And in corporeal things a thing that is not moved in some way itself cannot make another move.[13] (But although we do not put it into words, when we speak of movement we mean local movement, for we do not mean otherwise for now.)[14] And if you were to say to me that when a man throws a stone with his hand he does not move at all, I should say this is false; because even if the rest of his person does not move, the converse of this is not therefore true, for part of him <moves>, since his arm moves. And therefore the philosophers speak well, that 'Every natural agent is affected by acting.'[15] Thus to speak truly, a force or moving power has to produce a force on another and move it. And so each movement comes to have two principal aspects: the moving cause and the source and principle of the movement; and the mobile, in which as proper subject the movement is founded. Now this moving power is either intrinsic to the mobile or extrinsic. The intrinsic power is the cause of natural movement, as for example heaviness in the movement downwards of stones and levity in fire, which are the properties that follow from the form of stone and of fire. But the extrinsic power is the cause of violent movement, which we

guono le forme così del sasso come del fuoco. Ma la virtù estrinseca è cagione del movimento violento, del quale intendiamo noi hora di ragionare in buona parte. Non che per questo escludiamo dal nostro ragionamente il movimento naturale, ma perché tutti i movimenti mecanichi, ò la maggior parte, sono violenti et però nascono dalla virtù dell' agente ò della [S, 312v] forza sopra il mobile. Or perché il movente ha d' haver sempre proportione sopra il mobile, però diremo che secondo tal proportione verrà ad essere la velocità del movimento. Et perché ne movimenti violenti sempre il mobile resiste al movente, però secondo ancora la resistenza ch' è l' istessa quasi che la proportione detta di sopra, viene a cagionarsi la velocità del movimento. Ma questa resistenza ha da essere in alcuna proportione sensibile; perché quando fosse in proportione insensibile ò che si denominasse d' un numero grandissimo, non nascerebbe il movimento, si come ancora non nascerebbe il movimento dall' equalità delle proportioni.

A. Quantunque io habbia in mente le proportioni, però Vostra Altezza per mia sodisfattione sarà servita a darme alcun essempio?

P. Son contento. Dall' equalità non viene movimento questo, è quello che primo intendo di provare a Vostra Signoria. Mettiamo, che una forza sia determinata a cento[9] libre et si ponga a movere le cento libre, non è dubbio che le cento libre vengono a resistere col peso loro alla forza delle cento libre et, in modo restono rispetto all' equalità, che non ne può venir movimento, essendo tanto la resistenza quanto la forza. Il che Vostra Signoria può vederlo con l' essempio della bilancia, percioche se uno scudo ha da esser di giusto peso, fa che le bilancie non si piegano ne all' una parte ne all' altra; ma se sarà più che di peso traboccherà dalla sua parte et se meno dalla parte del peso. Chiamerò hora ó il peso ó lo scudo virtù movente resterà l' altro per quello che haverà da esser mosso. Se adonque la virtù movente è equale a quella del mobile, non si farà movimento, si come dall' essempio proposto ha veduto. Se la virtù movente è maggiore che quella che ha da esser mossa, il mobile si moverà, et sarà violentato dalla virtù movente. Ma per il contrario, se la virtù movente sarà meno potente della mobile, sarà la movente violentata dalla mobile [S, 313r] et in iscambio di far muovere sarà mossa. Et però hanno detto bene i filosofi, che *ab equali non provenit actio,* salvo se non volessimo dire, che l' attione fosse d' equalità, risposta d' alcuni puochi considerati, poiché per l' attione s' intende in questo caso il movimento d' alcuna sorte, il che, come habbiamo mostro, non viene dalla equalità.

9 cento] centro *S*

intend now mainly to discuss. Not that we are excluding natural movement from our discussion; but since all or most mechanical movements are violent, they therefore arise from the power of the agent or force [S, 312v] on the mobile.[16]

Now, because the mover must always have a ratio over the mobile, we say that the speed of the movement will come to be according to that ratio. And because in violent movements the mobile always resists the mover, the speed of the movement comes to be caused again in accordance with this resistance, which is almost the same as the ratio just mentioned. But this resistance must be in some perceptible ratio; because if it were in an imperceptible ratio or denominated by a very large number, movement would not arise; nor would movement arise from an equality of ratios.[17]

AN. Although I have the ratios in mind, for my satisfaction could you please give me an example?

PR. I would be happy to. First I intend to prove to you that movement does not come from equality. Let us suppose that a force is set at one hundred pounds, and one tries to move the hundred pounds, there is no doubt that the hundred pounds come to resist with their weight a force of a hundred pounds, and, as long as they remain in equality, that movement cannot occur, since the resistance is as much as the force. You can see this in the example of the balance, for if a *scudo*[18] is of the correct weight, the balance tips neither to one side nor the other; but if it is more than the counterweight it will tilt to its side and if less to the side of the counterweight. Now whether I call the counterweight or the *scudo* the moving power will depend on which causes the other to move. Then, if the moving power is equal to that of the mobile, there will not be movement, as you saw from the example. If the moving power is greater than what has to be moved, the mobile will be moved, and it will be forced by the moving power. But on the contrary, if the moving power is less powerful than the mobile, the mover will be forced by the mobile [S, 313r] and instead of making something else move it will be moved itself. And therefore the philosophers spoke well, that 'action does not arise from the equal,'[19] unless one should not even say that 'action' could be from equality, as some of little repute have quibbled, since by 'action' in this case is meant movement of some sort, which, as we have shown, does not come from equality.[20]

Appresso, la proportione dell' uno sopra dell' altro non haverà da essere denominata da numero grandissimo, perché a tal modo non avverà movimento, come sarebbe à dire da uno à mille, ò quattro milia, ò da uno à cento milia. Perché ogni volta che la forza superasse il mobile in alcuna di queste proportioni, non ne seguirebbe movimento di qualità. Anzi ne seguirebbe confusione, si come appare nelle navi, che vengon pinte dal vento, il quale s' è convenevole fa andar la nave di convenevole movimento ancora. Ma quando il vento supera di gran lunga la resistenza della nave, ne nasce la rotta di quella. Poiché i naviganti prima tagliano gli alberi et sostentano le vele, et doppo tagliano i cartelli, così di poppa come di prora, et solo ritengono il temone, il quale poi quando vien loro tolto[10] dal mare, è cagione della manifesta perdita della nave. Così ancora possiamo dire del movimento delle carrozze et delle altre cose tirate. Possiamo ancora applicar questo all' arteglierie di bronze lunghe come cannoni et colubrine, alle quali perché l' esperienza haveva dimostrato che le balle di pietra et di piombo non resistevano: queste perché la forza del fuoco delliquava prima che al destinato luogo venissero; et quelle per la leggerezza loro ó cadevano prima che venissero al proposto luogo ó derivavano dal camino. Hanno però applicate quelle di ferro, le quali resistono et al foco et alla forza della polvere. Dalla poca osservationi et intelligenza di questo principio sono avvenuti i molti ingegnieri de nostri et de tempi passati molti inconvenenti.

A. Et io saprei dirne alcuni. Da questa propo[S, 313v]sitione stimo che avvenga che molte volte un huomo volendo tirare un picciol sasso, con quanta più forza vi pone meno da se lo scaglia. Et io mi ricordo ch' essendo in Napoli et in Sicilia giovenetto haveva[11] preso un gran pezo di pomice[12], et lanciatolo con intentione di mandarlo lontanissimo da me, et m' avvedevo che mi cadeva poco longi da piedi con farmi dolere il braccio. Da qui veggo ancora la solutione del problema delle canne, perché quelle vengono à resistere alla forze de fanciuli et non à quella de gl' huomini.

P. Vostra Signoria dice bene, ma oltre alla resistenza i fanciulli v' usano una certa destreza, la quale è quella che più et meno lontano fa andarle.

A Io so benissimo che in tutte le forze vi bisogna la destreza, poiché nel giuoco delle canne et in quello de caroselli l' ho osservato benissimo, et insieme nel lanciar de dardi et delle zagaglie alla Moresca. Ma mi

10 tolto *corr. ex* tolte *S*
11 haveva] havea *S*
12 pomice] Roi nace *S*

Next, the ratio of the one over the other must not be denominated by a very large number, because in that case there would not be any movement – for example, by one to one thousand, or four thousand, or by one to one hundred thousand. Because whenever the force exceeds the mobile in one of these ratios there would not follow any orderly movement. In fact, there would follow confusion, as appears in ships that are pushed by the wind, which, if it is moderate, makes the ship also go with a moderate movement. But when the wind exceeds the resistance of the ship by a great deal, the ship will begin to founder. Then the sailors first shorten the masts and keep the sails on; afterwards they take in the sails both fore and aft, and they keep only the rudder, which, when it is taken away by the sea, is the cause of the clear loss of the ship.[21] We can say the same of the movements of carts and other things that are pulled. Again, we can apply this to long bronze artillery such as cannons and culverins, to which experience has shown that balls of stone or lead do not offer <the properly proportioned> resistance: lead balls because the force of the firing dissipates before they reach their target; and stone balls, which because of their lightness either fall before they reach the target or deviate from their course. Therefore balls of iron are used, which resist both the firing and the force of the powder. From few observations and little understanding of this principle many problems have been encountered by many engineers of our day and of previous times.[22]

AN. And I could certainly tell you some of them. This explanation [S, 313v] I think is the reason that often when a man wants to throw a small stone, the greater the force he throws it with the less distance it goes. And I remember as a boy in Naples and Sicily I would take a great piece of pumice stone and try to throw it very far from me, and I noticed that it fell only a short distance from my feet and hurt my arm.[23] From this I also see the solution of the problem of the reeds, because they can offer resistance to the force of boys but not to that of men.

PR. You speak well; but besides the resistance the boys use a certain skill that makes them go farther or less far.

AN. I know very well that with all forces skill is required, for in the game with reeds and in tournaments I have observed it very well, and also in throwing spears and the African assagai. But a doubt

viene un dubbio, ch' è: dove possa essere che la resistenza sia cagione de movimento?

P. Io mi protesto che con Vostra Signoria non ragiono ne con quelle divisioni ne con quelle distintioni ne con quelle sutterfugii che ragionerei con un filosofo, con il quale ver rei sino a considerare la iacitura delle lettere nelle parole non che altro. Et però ragionando con un cavaliero soldato ragionerò con principii che più sono manifesti al senso, lasciando il resto alle persone suoperate non per saper ma per contender vaghe. Domanderò a Vostra Signoria volendo ella levare verbi gracia da terra quella colubrinetta che quivi giace, ch' è quello che fa, che non possa levarla facilissimamente?

A. È la gravità, su la quale per propria natura la tien all' ingiù.

P. Bene ha detto. Pigli hora questa balla di ferro; volendola tirare molto lontano, ch' è quello che non lassa farglielo?

A. Parimente la gravità sua, la quale fa che ella tenti subito di discendere, uscendomi dalle mani non havendo tanta forza di superar quella.

P. Et volendola tirare, bisogna che sforzi se stessa per vincere la violenza della balla, et così [S, 314r] viene ad accrescer la sua forza per sforzare la resistenza della balla. Or quando la gravità non fosse tanta che potesse cagionare la moltiplicatione della forza con resistere a quella (anzi senza far punto di resistenza cedesse subbito), in essa non s' imprimerebbe la virtù movente. Ne altresi potrebbe imprimersi quantunque ella in alcun modo si accrescesse, quando la resistenza fosse tanta che di gran lunga superasse la virtù. La cosa adonque ch' ha da esser mossa, ha da haver due conditioni, delle quali mancandoni una non si farà il movimento. La prima è ch' ella sia atta a cedere alla forza, perché come non cedesse non nascerebbe il movimento; la seconda, che sia atta à resistere al movente. Ogni volta che mancasse d' una di queste conditioni non si potrebbe fare l' attione mobile, perché se resistesse solo et non cedesse[13], non si moverebbe; si cedesse solo et non resistesse, non verrebbe pinta inanti. È adonque la resistenza cagione di movimento, in quanto che resistendo et cedendo fa che la virtù movente con più vehementia s' imprime nel mobile. Il tutto Vostra Signoria può haverlo dall' arteglierie, che maggior percossa fanno dove è maggiore la resistenza. Da dove è che si sono messi in uso i terrapieni et le muraglie di pietra cotta; et da dove è ancora, per quanto ho inteso, che le mura di Napoli sono atte a resistere all' arteglierie per esser d' una pietra assai più tenera delle pietre cotte.

13 cedesse] cadesse *S*

occurs to me, which is, how can it be that resistance is the cause of movement?

PR. I profess that with you I shall reason neither with those divisions nor with those distinctions nor with those subterfuges that I would use with a philosopher, with whom in truth I would consider just the tossing of letters into words and nothing else. And therefore, reasoning with a cavalry officer, I shall reason with principles that are more manifest to sense, leaving the rest to retired persons not for knowledge but for vague contention. I ask you: if you wanted to lift from the earth that culverin that lies here, for example, what is it that makes it not easily lifted?

AN. It is weight, which through its proper nature holds it down.

PR. You spoke well. Now take this ball of iron; if you wanted to throw it very far, what would not let you do it?

AN. Similarly its weight, which makes it descend immediately, since when it leaves my hand it does not have enough force to overcome it.

PR. And in order to throw it, it is necessary to overcome the violence of the ball, and so [S, 314r] the hand tends to increase its own force to counteract the resistance of the ball. Now if the heaviness were not enough to cause the multiplication of the force by resisting it (in fact, without any resistance at all the mobile would yield at once to the force), the moving power would not be impressed on the mobile. Nor again could it be impressed, even if it were somewhat increased, when the resistance is so much that it exceeds the power by a long way. The thing that is to be moved, then, must have two conditions, of which if one is missing there will not be movement. The first is that it must be able to yield to the force, because, if it did not yield, movement would not arise; the second is that it must be able to resist the mover. Any time one of these conditions is lacking the mobile cannot be set in motion: because if it only resisted and did not yield, it would not be moved; and if it only yielded and did not resist, it would not remain in motion.[24] And thus resistance is the cause of movement, in so far as resisting and yielding impress the moving power on the mobile with more vehemence. You can see it all in artillery, which makes the greater impact where the resistance is greater. Whence it is that ramparts and walls of terracotta are used; and whence it is also, so far as I have heard, that the walls of Naples are able to resist artillery by being of a stone even less dense than terracotta.[25]

Or perché nelle cose materiali non si può descendere all' ultima proportione, da qui è che noi non possiamo dire un sasso gittato[14] da un braccio con una fromba moversi nel più veloce modo ch' egli possa, essendo che un' altro può farlo col medesimo stromento movere più velocemente et un altro meno. Et che nell' esser il sasso di tanto peso, possa trovarsi una virtù che possa farlo movere col più veloce movimento che mover si possa, si può così provare. Se quel sasso rispetto ad una minore virtù non viene à moversi per rendersi gravissimo [S, 314v] et rispetto ad una maggior virtù non viene à moversi per esser leggierissimo, adonque tra queste due virtù sarà una virtù di mezo, alla quale il sasso sarà accommodato benissimo, la quale è però difficile da ritrovare nelle cose materiali ma non però impossibile, et noi appresso tenteremo di ritrovarla.

A. Bel discorso è stato questo, et palpabile. Ma è un[15] problema in Aristotele dove s' assegni la cagione di questo?

P. Parmi haver detto a Vostra Signoria di si, et è il suo 34ª delle *Meccaniche*, et ne rende la cagione detta et oltre ne rende poi un' altra della aria. «Perché – dice egli – le cose molto grandi non posson lanciarsi molto lontane, ne altresi le molto picciole?» Doppo d' haver detto le ragioni dette di sopra, soggiugne et dice: «È forse perché le picciole non possono per la leggereza et piccioleza loro pingere molto l' aria, essendo che tanto si muovono le cose quanto l' aria vien mosso. Et però et le picciole et le grandi restano quasi immobili, queste come quelle che non posson moversi, et quelle non posson movere.»

A. Et che ha a fare l' aria[16] con le cose che si muoveno?

P. Ha a fare molto; anzi fa opinione d' alcuni che le cose gettate ricevesserò la continuatione del movimento loro dall' aria mosso.

A. Mi ricordo haver sentito disputare et conchiudere che la cagione del movimento delle cose gettate veniva ad essere la virtù movente, la quale, come Vostra Altezza ha detto, s' imprime nel mobile et così lo pinge innanzi.

P. Et questa questione ancora propone Aristotele in più luoghi et particularmente nelle *Meccaniche* nella 32ª et 33ª questione. Ma accioche Vostra Signoria sia informato del fatto, ha da sapere che 'l grave che si muove, et così il leggiero, ó si muove secondo la propria natura sua ó si

14 *ante* gittato *scrib. et del.* è *S*
15 un] uui *S*
16 aria] una *S*

Now because with material things one cannot descend to the ultimate ratio, we cannot say that a stone thrown by an arm with a sling is moved as swiftly as possible, since someone else with the same instrument could make it move more swiftly and someone else again less swiftly. And that a power can be found that can make a stone of a given weight move with an even swifter movement can be proved thus. If that stone in relation to a smaller power tends not to move by being very heavy, [S, 314v] and in relation to a larger power tends not to move by being very light, then between these two powers there will be a mean power, to which the stone will be best accommodated, but which is difficult but not impossible to find in material things; and later we shall try to find it.[26]

AN. This was a elegant and sensible explanation. But is there a problem in Aristotle where he gives the cause of this?

PR. Indeed there is, and it is the thirty-fourth question in the *Mechanical Problems*, where he gives for it the cause we just discussed, and then in addition gives another involving air. 'Why,' he asks, 'can very large things not be thrown very far, nor also very small things?' After giving the reasons mentioned above, he adds the following: 'It is perhaps because small things through their lightness and smallness cannot push much air, since the more things are moved the more the air tends to be moved. And so both the little and the great stay almost immobile, these because they cannot be moved, and those because they cannot move' <the air>.[27]

AN. And what does the air have to do with things that are moved?

PR. It has much to do with them; in fact it is the opinion of some that things thrown receive the continuation of their movement from the moved air.

AN. I remember having heard this disputed and the conclusion was that the cause of the movement of things thrown is the moving power, which, as you have said, impresses itself on the mobile and thus pushes it forward.[28]

PR. And this question Aristotle again poses in many places, particularly in the *Mechanical Problems*, questions 32 and 33.[29] But to be fully apprised of the facts, you should know that both heavy and light bodies are moved either according to their own proper natures or by violence. We say that a stone is moved according to its nature when it

muove per violenza. Diremo un sasso moversi secondo la natura sua quando descende verso il centro del mondo per linea perpendicolare Et così il leggiero diremo moversi secondo la sua natura quando anderà all' insù drittissimamente et per linea perpendicolare, che è [S, 315r] per l' asse dell' orizonte.

<Le velocità dei corpi cadenti>

[G, 3r] Or il grave, movendosi naturalmente, può muoversi con maggiore et con minore velocità rispetto al mezo, poiché per un mezo più sottile si muove con maggior velocità et per un mezo più grosso con meno. Il che tutto può Vostra Signoria intenderlo benissimo con le cose che si muovono all' ingiù per l' acqua et con quelle per l' aria. La dove se Vostra Signoria piglierà una profondità d' acqua di 100 passi, et vi lasciarà andare un grave et osservarà il tempo che consumerà à toccare il fondo et noterallo da parte; et di nuovo piglierà un' alteza di 100 passi parimente, et vi lascierà andare un grave del peso, sostanza, et figura dell' altro, et terrà conto del tempo che consumerà nel venire al basso, troverà questo tempo essere molto minore dell' altro.

A. Perché vuole Vostra Altezza che 'l grave sia della stessa sostanza, peso, et figura dell' altro?

P. Per levar le cagioni da dubitare.

A. Et che dubbio può nascere intorno à questo?

P.[17] Grandissimo. [G, 3v] Percioche Aristotele ha dato cagione da dubitare, dicendo che per uno stesso mezo la velocità delle cose che si muovono per movimento naturale, essendo della stessa natura et figura, è si come la potenza loro. Ciò è, se[18] dalla[19] cima d' un' alta torre noi lascieremo venir giù due palle, l' una di piombo di 20 libre et l' altra parimente di piombo di una libra, che 'l movimento della maggiore sarà vinti volte più veloce di quello della minore.

A. Questo mi pare assai ragionevole; anzi, quando mi fosse domandato per principio, lo concederei.

P. Vostra Signoria s' inganerebbe; anzi, vengono tutti in uno[20] stesso tempo, et di ció se n' è fatta la pruova non una volta ma molte. Ma che è di più[21], che una balla di legno, ó più ó men grande d' una di piombo, lasciata venir giù d' una stessa alteza nello stesso tempo con quella di

17 P. *corr. ex* Gr *S*
18 se *ins. S*
19 dalla *corr. ex* della *S*
20 *ante* uno *scrib. et del.* in *S*
21 che è di più] che più *S*; v' è di più *G*

descends towards the centre of the world along a vertical line. And similarly we say that a light body is moved according to its nature when it goes straight up along a line perpendicular [S, 315r] to the arc of the horizon.

<The Speeds of Falling Bodies>

Now a heavy body, being moved naturally, can be moved with greater or lesser speed depending on the medium, since through a rarer medium it is moved with greater speed and through a denser with less.[30] All of this you can understand very well with things that fall through water and those that fall through air. Whence if you took a depth of water of a hundred paces, and released in it a heavy body and observed the time it took to touch the bottom and made a note of it on the one hand; and again, if you took a height <in the air> also of a hundred paces, and let go from it a heavy body of the same weight, substance, and shape as the other, and measured the time it took to reach the bottom, you would find this time to be much less than the first.

AN. Why do you want the heavy body to be of the same substance, weight, and figure as the other?

PR. To remove the causes of doubt.

AN. And what doubts can arise concerning this?

PR. The greatest, because Aristotle gave rise to doubts by saying that through one and the same medium the speed of things that are moved in natural movement, being of the same nature and shape, is as their powers. That is, if we were to let fall from the top of a tall tower two balls, one of twenty pounds of lead and the other of one pound, also of lead, that the movement of the larger would be twenty times faster than that of the smaller.[31]

AN. This seems sufficiently reasonable to me; in fact, if I were asked I would grant it as a principle.

PR. You would be mistaken; in fact, both arrive at one and the same time, even if the test were done not once but many times. But what is more, a ball of wood, either larger or smaller than one of lead, let fall from the same height at the same time as the lead ball, would descend and touch the earth or ground at the same moment in time.

piombo, descendono ò toccano[22] la terra ò 'l suolo nello stesso momento di tempo.

A. Se l' Altezza Vostra non mi dicesse d' haverne fatta la prova, io nol crederei. Et come si può salvare Aristotele?

P. Molti si sono [S, 315v] sforzati di salvarlo diversamente, ma in fatti mal si può salvare. Anzi, per dir à Vostra Signoria il [G, 4r] tutto, io credei un giorno d' haver trovato il modo di salvarlo, ma poi pensando meglio al fatto, così non fù.

A. Tuttavia non può essere che 'l modo non sia ingenioso et arguto, et però Vostra Altezza sia servita à dirlo.

P. Per compiacerla la dirò, ma prima dichiarerò alcuni principii che mi bisognano. È chiara cosa appresso à filosofi[23] che quanto più un grave si muove per proprio movimento, come il sasso col descendere, tanto più venghi velocitandosi. La dove chi presupponesse uno spacio infinito, infinita sarebbe per quello la velocità del grave. Se adonque presupponessimo che nel concavo della luna fosse un grandissimo sasso, prima che fosse nella superficie della terra si sarebbe fatto di movimento molto veloce. Può di questa velocità Vostra Signoria certificarsene oltre l' autorità de'[24] filosofi in questo modo. Potrà pigliare una balla ó di sasso ó di piombo ó di ferro ó d' altra materia grave et lasciar venir giù questa balla da due diverse alteze, la quale percuota in due resistenti d' ugual natura; vedrà che quella che verrà dal luogo più alto farà maggior effetto nel resistente, che quella che verrà [G, 4r] dalla minore alteza, et non essendo la stessa cosa cresciuta di peso. Adonque haverà da dire il maggior effetto venire dalla maggiore velocità? Appresso stante questo principio, se noi faremo d' una stessa alteza venir due palle di disugual grandeza et siano della stessa materia, è manifesto che la maggiore nello stesso resistente farà maggior effetto che la minore. Adonque sarà venuta con maggior velocità che la minore; adonque non si muovono con ugual velocità, che è quello[25] che si vuole.

A. Ho inteso la ragione di Vostra Altezza et in vero par che possa salvar Aristotele. Ne saprei per hora trovarvi l' inganno se non vi pensasi su.

P. L' inganno è facile da scoprire, poiché la maggior percossa della maggior balla non nasce dalla velocità del movimento, essendo [S, 316r]

22 ò toccano] o trovano *G*
23 à filosofi *om. G*
24 de'] de *S*
25 *ante* quello *scrib. et del.* quando *S*

AN. If you had not told me that you had made the test I would not believe it.[32] So how can Aristotle be saved?

PR. Many have [S, 315v] striven to save him in various ways, but in fact he cannot very well be saved. Indeed, to tell you the truth, I believed one day I had found the way to save him, but then thinking more about it, I was wrong.

AN. Nevertheless, it cannot be that your way was not ingenious and clever, and so it would do you credit to tell it.

PR. To please you I shall tell it, but first let me explain some principles that I shall need. It is well known among philosophers that the more a heavy body is moved in its proper movement, as a stone is in descending, the more it tends to go faster. Whence for those who posit an infinite space, the speed of heavy bodies falling through it would increase to infinity.[33] If we therefore suppose that at the sphere of the moon there were a huge stone, before it came to the surface of the earth it would have to move very swiftly.[34] You can confirm this speed (beyond the authority of the philosophers) in this way. Take a ball either of stone or lead or iron or some other heavy material and drop this ball from two different heights and let it strike two equal resistances. You would see that when it falls from the higher place it has a greater impact on the resistance than when it falls from the lesser height, without increasing in weight. Would you not have to say then that the greater impact comes from the greater speed?[35] Next, having established this principle, if we drop from the same height two balls of unequal size and of the same material, it is apparent that the larger will make a greater impact on the same resistance than the smaller. Therefore it will arrive with greater speed than the smaller; thus it will not be moved with an equal speed, which is what was desired.[36]

AN. I have understood your reasoning and it seems that it could indeed save Aristotle. I would not know right off how to find the mistake without thinking about it.

PR. The mistake is easy to discover, because the greater impact of the greater ball does not arise from the speed of the movement – since [S, 316r] sense observes that the movements are equal – but it arises from the weight, which one can prove as follows. Let us drop from high

che 'l senso osserva esser il movimento equale, ma nasce dal peso, il che si può provare così. Lasciamo venir da alto et da due diverse distanze due palle della medesima materia ma[26] di disugual peso, et venghi la minore dalla maggior alteza, la quale ecceda la minore nel tripplo ò nel quadruplo, [G, 5r] ò facciamo che la minore di due onze venghi d' una alteza di 100 passi et la maggiore di due ò tre libre venghi non più da alto che da quattro ò di cinque passi: quale crede Vostra Signoria che nello stesso resistente farà maggior effetto et percossa?

A. Et chi dubita che la maggiore? – et così dimostra l' esperienza.

P. Et ciò di dove è se non dal maggior peso? et con tutto ció, con maggior velocità descende la minore poiché da maggior alteza viene. Essi poi sforzato Girolamo Cardano nel libro suo *Delle proportioni* di dimostrar che due palle di disugual grandeza messe in pari[27] alteza sieno pervenir[28] giù nello stesso tempo. Ma perché la demostratione sua non mi piace intieramente, io lascio di dirla à Vostra Signoria.

A. Anzi voglio supplicare Vostra Altezza che me la[29] dica per vedere l' errore d' un' huomo così famoso.

P. Io non voglio dire che sia errore, ma ho solo detto che non mi piace, et dirola per sodisfare à Vostra Signoria. Egli dice: sieno due palle,[30] A maggiore et B minore, et il diametro di A sia di tre[31] palmi, verbi gracia, ò qual altra misura si voglia, et quello di B uno della stessa mesura, et sieno della medesima materia et sieno messe con ugual distanza da CD, il quale sia il piano dove sieno per dare [*vide figuram XII*]. Dico che, lasciare andare nello stesso tempo, che parimente nello stesso daranno nel piano CD. Perché il diametro del corpo A è triplo al[32] diametro del corpo B, adonque il corpo A al corpo B sarà come ventisette ad uno, poiché le sfere hanno la proportione tra di loro che i cubi dei loro diametri, per l' ultima del duodecimo d' Euclide. Adonque [S, 316v] la gravità di A alla gravità di B è come di 27 ad uno. Ma perché ogni peso nel descender suo condensa l' aria in quel grado ch' egli pesa, come l' aria sotto A è 27 volte più densa che l' aria sotto B, però il peso

26 ma] una *S; corr. ex* una *G*
27 pari] più *S G*
28 pervenir] per venir *S G*
29 la] le *G*
30 palle *corr. ex* balle *S*
31 di tre] BC *S*; di 3 *corr. ex* BC *G*
32 al] ad *S G*

up and from two different heights two balls of the same material but of unequal weight, and let the smaller fall from the greater height, which exceeds the smaller height by three or four times; and let us suppose that the smaller (of two ounces) falls from height of a hundred paces and the larger (of two or three pounds) comes from no higher than four or five paces: which do you believe will make the greatest effect and impact on the same resistance?

AN. And who doubts that the larger would? – and experience demonstrates it.

PR. And so where does it come from if not from the greater weight? And all this being so, the smaller descends with greater speed because it comes from a greater height. These considerations then forced Girolamo Cardano in his *Opus novum de proportionibus* to demonstrate that two balls of unequal size fall from a great height at the same time. But because his demonstration does not entirely satisfy me, I shall omit giving it to you.

AN. Yet I beg you to tell me so I might see the error of so famous a man.

PR. I would not want to call it an error, but I said only that it does not satisfy me, and I shall give it to gratify you. He said: let there be two balls, A the larger and B the smaller, and let the diameter of A be three spans, for example, or any other unit you like, and that of B one of the same units, and let them be of the same material and let them be put at an equal distance from CD, which is the plane where they will hit (see figure XII). I say that, if they are dropped at the same time, they will hit the plane CD at the same time. For since the diameter of body A is triple the diameter of body B, body A will then be to body B as twenty-seven to one, since spheres have ratios between them as the cube of their diameters (by Euclid, *Elements* 12.18).[37] Therefore [S, 316v] the heaviness of A to the heaviness of B is as twenty-seven to one. But because every weight in falling compresses the air to the degree that it weighs, as the air under A is twenty-seven times denser than the air under B, therefore weight A, having to pass through denser air, necessarily has more difficulty in its descent. Thus, since the ratio of A to B is as twenty-seven to

A, havendo da passare aria più densa[33], forza è che più peni nel descender suo. Adonque, essendo la proportione di A à B come 27 ad uno et tale essendo la potenza di A à B, seguirebbe che quando non havesse impedimento che si dovesse movere nella [G, 6r] velocità di 27 ad uno. Ma[34] l' impedimento di A all' impedimento di B è come 27 ad uno; adonque equale è l' impedimento alla potenza, et però seguirà che 'l movimento loro debba esser in ugual tempo.

A. Se 'l Cardano la mette così facile et chiara come Vostra Altezza l' ha detto, à me pare una bella demostratione; ne saprei per quel ch' io me ne intenda, dire se non che fosse intieramente et per ogni parte bella.

P. À me piace più adesso che l' ho detto à Vostra Signoria che quando la lessi appresso dell' autore. Et quel che à me non piaceva era quella densità, perché non son ben capace che l' aria si condensi secondo il peso. Che si condensi ancora[35] si potrebbe dubitare. Ma concedendo che l' aria si condensi, et che si condensi secondo il peso la demostratione corre benissima, et è bella et ingeniosa. Quanto al condensarsi dell' aria, molti par che lo concedano et particolarmente nella cosa de'[36] movimenti, perché, quando non si concedesse tal condensatione, saremo sforzati à concedere il vacuo, [G, 6v] cosa tanto odiosa alla natura, la quale più presto comporta che le cose gravi ascendono che admettere quello.

A. Et come potrà Vostra Altezza mostrare che la natura admette più tosto che le cose gravi ascendano che 'l vacuo?

33 densa] denso *S G*
34 Ma] tra *S G*
35 ancora] encora *G*
36 de'] de *S*

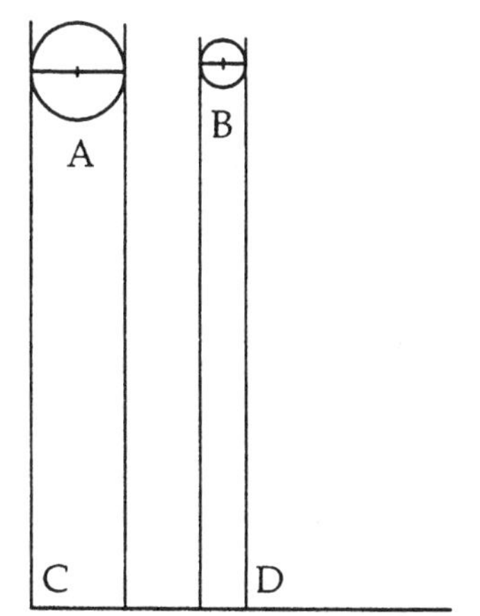

Figure XII

one and so is the power of A to B, it follows that when there is no resistance they must be moved at speeds of twenty-seven to one. But the resistance of A to the resistance of B is as twenty-seven to one; therefore the resistance is equal to the power, and thus it would follow that their movements must be in equal times.[38]

AN. If Cardano put it so simply and clearly as you told it, it seems a sound demonstration to me; from what I heard I would not be able to say that it is not entirely and in every part sound.

PR. It satisfies me more now when I tell it to you than when I read it in the author. And what did not satisfy me was the density; because it is not at all certain that air is condensed in proportion to the weight. That it is condensed at all could even be doubted. But assuming that the air is condensed and that it is condensed in proportion to the weight, the demonstration runs very well, and it is sound and clever. As for the condensing of the air, many also concede it and particularly in the case of movements, because, if such condensation were not conceded, we would be forced to concede the vacuum, so odious to nature, than which to admit one more readily allows that heavy things ascend.[39]

AN. And how will you be able to show that nature admits more readily that heavy things ascend than the vacuum?

P. In molti modi potrei mostrarlo à Vostra Signoria, ma perché questo non è il suo luogo, però sarà bene soprasedere alquanto.

<*La causa perché il grave, quanto più discende, tanto più viene velocitandosi*>

A. Soprasederò, ma vorrei intendere se fosse possibile dimostrare perché il grave, quanto più descende, tanto più viene velocitandosi – perché mi pare haver sentito dire, non so che di luogo.

P. Dirò à Vostra Signoria [S, 317r] intorno à ciò molte sono state le opinioni, ma le famose sono l' una del luogo, l' altra del movimento, la terza rispetto al mezo[37], et questa par ch' habbia del demostrativo.

Quanto al luogo, molti hanno detto che' l luogo è cagione della velocità del grave et così del leve, dicendo che 'l grave appetisce d' andare al luogo suo, et però quanto più à quello s' appressa, tanto più si velocita per arrivar presto à quello[38]. Il che non pare che possa esser [G, 7r] vero, essendo che quando così fosse, nel grave verrebbe ad essere una virtù cognoscitiva[39], cosa fuori del ragionevole.

L' altra è del movimento, percioche essendo[40] il movimento l' atto del mobile, et l' atto essendo la perfettione della cosa, adonque quando il grave si muove, è nella sua perfettione. Ma chi è già in atto siegue l' operatione che da quell' atto viene con più facilità nell' ultimo, che nel principio ò nel mezo. Adonque cominciando il grave à moversi, non si muove con quella facilità[41], che fa doppo che si sarà mosso per alquanto di spacio, essendo che viene alterandosi di mano in mano, et però quanto più si moverà, con tanto più facilità verrà à moversi et per conseguente con tanto più velocità. Da dove è che con più velocità si muove nel fine che nel principio ò nel mezo.

A. Questa mi pare demostratione, et nella quale non n' è cosa alcuna da negare. Et parmi simile alle ragioni che si dicono nelle morali, che, come l' huomo ha acquistato [G, 7v] l' habito delle virtù, le fa senza fatica, et quanto più opera, tanto più apprende. È simile ancora à quello che diciamo dell' intendere, che l' intelletto non s' affattica nell' intendere. Et son certo che se Franceschino, che suona l' organo di santa Barbara, non

37 mezo] modo *G*
38 quello] quella *S*
39 cognoscitiva] conosciuta *G*
40 *ante* essendo *scrib. et del.* eg *S*
41 facilità] facultà *S G*

PR. I could show you in many ways, but because this is not the place, it were best to pass over it.

<The Cause of the Acceleration of Falling Bodies>

AN. I shall let it pass, but I should like to hear if it is possible to demonstrate why the more a heavy body falls the more it tends to go faster – because it seems to me I have heard it said, but I do not know where.

PR. I must tell you [S, 317r] that there have been many opinions about this, but the famous ones are the one involving place, the second involving movement, and the third with respect to the medium, and this last appears to have demonstrative force.[40]

As for place, many have said that place is the cause of the speed of heavy and light bodies, saying that the heavy body seeks to go to its place, and therefore the more it nears it, the more it speeds up to arrive in it sooner.[41] Which cannot possibly be true, since if it were so, the heavy body would have to possess a cognitive power, a thing beyond the reasonable.[42]

The second concerns movement, because movement, being the act of the mobile, and act being the perfection of the thing, it follows that when a heavy body is moved, it is in its perfection; but what is already in act continues the work that arises from this act with more facility at the end than at the beginning or in the middle. Therefore, when it first begins to be moved, a heavy body is not moved with the facility that it has after it has been moved through some distance, since it comes to be changed as it goes along; and therefore the more it is moved, with the more facility it will be moved and consequently with more speed. Whence it is that it is moved with more speed at the end than at the beginning or in the middle.

AN. This seems demonstrative to me, and there is nothing in it to deny. And it seems similar to me to the reasons one gives in ethics, that when a man has acquired the habit of virtues, he does them without effort, and the more he does the more he acquires. Again it is similar to what we say of understanding, that the intellect does not tire from understanding. And I am certain that if Francescino, who plays the organ of Santa Barbara,[43] did not feel the physical labour, that the more

sentisse il travaglio del corpo, che quanto più sonasse, tanto più suonarebbe; ma è forza che all' ultimo le membra s' affattichino. Ciò non può avvenire al grave poiché le parti sue non s' affatticano nel descendere per esser cosa inanimata; et però come Vostra Altezza ha detto, et bene, più et più si velocita dall' attirarsi più et più col descender, [S, 317v] et io quanto à me mi contenterei[42] di questa sola ragione. Ma se Vostra Altezza per suo contento vuol dir l' altra, io l' udirò voluntieri.

P. Poiché s' è nominata, è ben dirla, perché acqueta non meno della pur hora detta, et è messa dal Cardano: s' ho[43] da provare che 'l grave descendendo, quanto più descende, tanto più si fa veloce nel movimento suo, essendo tale il movimento suo naturale. Si presuppone con verità che l' aria sia l' impedimento al movimento del grave; poiché, come prova Aristotele, quando dal concavo della luna infino al centro dell' universo non fosse corpo di sorta[44] alcuna ò fosse il luogo vacuo, il movimento si farebbe in instante. Ma quanto più l' aria è presso, tanto più si condensa, et quanto più è condensato, tanto più resiste al movimento del grave. Adonque, mentre il grave si muove, quanto più è lontano dal luogo dove ha da andare, tanto più ha d' impedimento poiché tanto più aria ha da passare et per conseguente condensato dal peso del grave, et però più tardo sarà di movimento. Ma quanto più descenderà,

42 contenterei] contenteria *S G*
43 s' ho] s' o *S*; s' a *G*
44 sorta] sorte *S G*

he plays the more he would play; but it is necessary that ultimately the limbs will tire. This cannot happen to a heavy body because, being an inanimate thing, its parts do not get tired by falling; and therefore, as you have said – and said well – it goes faster and faster by being pulled more and more as it falls [S, 317v], and as far as I am concerned I shall be content with this reason alone. But if you for your own satisfaction want to give the other reason, I shall gladly listen.

PR. Because it has been mentioned, it is well to give it, since it is no less conclusive than what was just said, and it was proposed by Cardano: we are to prove that when a heavy body falls, the more it falls the more swiftly it moves, this movement being its natural one. We assume – with truth – that the air is an impediment to the movement of the heavy body; then, as Aristotle proves, if from the sphere of the moon to the centre of the universe [i.e., the centre of the earth] there is no body of any kind or there is empty space, the movement would be in an instant.[44] But the more the air is pressed the more it is condensed, and the more it is condensed the more it resists the movement of the heavy body. Thus, while the heavy body is moving, the farther it is from the place where it has to go, the more resistance it has because the more air it must pass through (which consequently is condensed by the weight of the heavy body), and therefore the slower the movement will be. But the

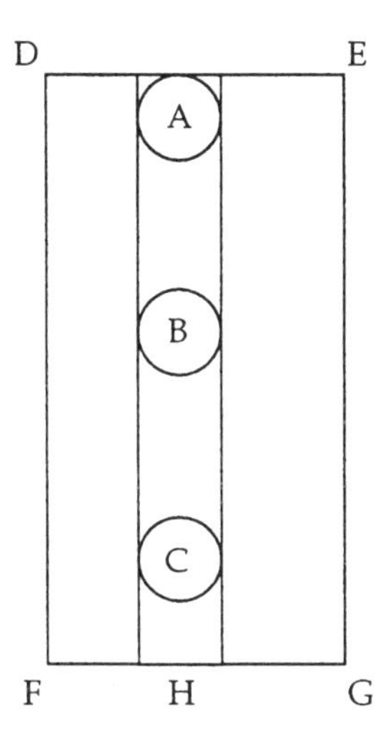

Figure XIII

tanto meno haverà d' impedimento et però più veloce sarà. Giugniamo à questo che se noi intenderemo il grave A nel concavo della luna, inteso per DE [*vide figuram XIII*], nel dispiecarsi da quel luogo, intendendo il piano della terra essere FG [G, 8v], haverà il cilindro ABH d' aria densa[45] da passare, et doppo lui non sarà aria che le succeda. Ma quando il corpo A sarà venuto nel B, oltre ch' haverà solo il cilindro BH da passare, che resisterà meno di quel che faceva ABH, haverà l' aria AB, che[46], succedendole per ragione del vacuo, verrà con l' impulso suo à velocitare il movimento del grave. Et per conseguente, quanto più descenderà, tanto maggior impulso haverà et minore resistenza, [S, 318r] et però maggiore sarà sempre la sua velocità. La dove, quando sarà in C, sarà di maggior velocità, che quando sarà in B, per l' allegate ragioni. Et così è vero che, quanto più il grave descenderà, con tanto maggior velocità si moverà.

45 densa] denso S G
46 che] et *S G*

more it descends the less resistance it will have and therefore the faster it will be. Let us add to this that, if we suppose the heavy body A to be at the sphere of the moon designated by DE (see figure XIII), in falling from there to the surface of the earth FG, A would have to pass through the cylinder ABH of dense air, and above it [i.e., above DE] there would be no air that could follow it. But when the body A has come to B, besides having only the cylinder BH to pass through, which will resist less than ABH, it has the air AB, which, following it because of the vacuum, tends with its impulse to speed up the movement of the heavy body. Consequently, the more it falls the more impulse it will have and the less resistance, and thus its speed will be ever greater. Therefore, when it is at C, it will have more speed than when it is at B, for the reason given. And thus it is true that the more a heavy body falls, with so much more speed is it moved.[45]

NOTES TO THE EDITION

The First Day

1 Moletti is perhaps referring here to Tartaglia's analysis of projectile motion into circular and straight parts and his application of this to cannonballs in his *Nova scientia* (Venice, 1537), translated in part in Stillman Drake and I.E. Drabkin, *Mechanics in Sixteenth-Century Italy* (Madision: Univ. Wisconsin Press, 1969), 63–97. In book 1 of his later *Quesiti et inventioni diverse,* Tartaglia goes further and tries to explain the force of impact of projectiles through the circular movement of the lever and balance (Niccolò Tartaglia, *Quesiti et inventioni diverse* [Venice, 1546; 1554; rpt. in facsimile, Brescia: Ateneo, 1959], ff. 5, 17v–19v); see the Introduction, 22–3 above.

2 Tartaglia nowhere discusses the pulley.

3 Cf. Tartaglia, *Quesiti et inventioni diverse,* book 6: 'Sopra il modo di fortificar le Città rispetto alla forma' (ff. 64–77).

4 Moletti uses the same examples in his earlier *Discourse on How to Study Mathematics:* 'Mi pare hoggimai tempo di passar alla mecanica, scienza tanto desiderata da molti e tanto alla vita humana necessaria. Ma è da sapere che per scienza mechanica, non intendo l' arti mechanice, come molti falsamente hanno stimato estimano, come l' arte del sarto, del muratore, del calzolaio, et come l' arte del fabro, del marangone, della lana, e tali; ma intendo quella scienza, che con dimostrationi mathematice aiuta la vita humana, superando le difficultà della natura et operando molte volte contra le inclinationi di quella: come far andar con arte l' acqua contra 'l corso suo, ascendere le cose gravi, descendere le leggieri, e tali. Il che tutto si fa con arte grande, et come ancora che una picciola forza facci operatione equale ad una grande e tali' (*Discorso sulla matematica,* Milan, Biblioteca Ambrosiana MS. S 103 sup., f. 155v).

5 Perhaps this is G.T. Scala, the author of *Architettura militare* (Venice, Correr, MS. P.D.C. 255; my thanks to Nicholas Adams for this reference), who is also perhaps the Giovanni Scala, author of *Delle fortificazioni* (Rome, 1596), listed in Horst de la Croix, 'The Literature on Fortification in Renaissance Italy,' *Technology and Culture* 4 (1963): 30–50, at 49). By 'senza lettere' Moletti probably meant what Leonardo did when speaking of himself: that he hardly knew how to read and write Latin.

6 See Vitruvius, *I dieci libri dell'Architettura di Vitruvio,* book 7, *Praefatio,* tr. Danielo Barbaro (Venice, 1556), 182; cf. *De architectura libri X,* ed. and tr. Frank Granger, 2 vols., Loeb Classical Library (London: Heineman; Cam-

bridge, Mass.: Harvard Univ. Press, 1956), 2:74–5; and cf. 'Post hos scriptores qui di machinationibus scripserunt, inveniuntur *Collectoria* Pappi Alexandrini, á clarissimo Federico Commandino latina facta, sed non dum tipis data. Etiam reperiuntur Heronis Alexandrini liber *De spirit<u>alibus*, et de his que per se moventur [= *Automata*], quorum tantum prior tipis latinus facto, ab eodem Commandino datus est. Nonnulla de machinamentis Athenei, sed graeca et manuscripta. Reperitur Hero *De machinis bellicis*, sed non ille Alexandrinus, verum illo multo posterior, et iam est impraessus, quem latinum fecit doctissimus Patritius Venetus Franciscus Barocius. Extat et libellus Philosophi quem exponere decrevimus, cuius libelli extat paraphrasis Georgi Pachymerii, latina lingua donata a Germano. Apud latinos non memini me videsse nisi ea quae quod Vitruvium reperiuntur, quae certe non pauca sunt, nam sunt principia ad hanc scientiam' (Giuseppe Moletti, *In librum Mechanicorum Aristotelis expositio*, Biblioteca Ambrosiana MS. S 100 sup., f. 184v); the last sentence here seems to convey the sense of the garbled Italian.

7 This is a close paraphase of the opening paragraph of the *Mechanical Problems:* 'Miraculo sunt ea quidem quae natura contingunt quorum ignorantur causae, illa vero que praeter naturam quaecunque ad hominum utilitatem arte fiunt. In multis enim natura ei quod nobis usui esse potest contrarium facit. Natura etenim eundem semper habet modum et simpliciter, utile autem multifariam commutatur. Quando igitur quippiam praeter naturam oportuerit facere difficultate sua haesitationem praestat arteque indiget. Quamobrem eam artis partem que huiusmodi succurrit difficultatibus mechanicam appellamus. Quemadmodum enim Antipho scribit poeta sic se res habet: Arte enim superamus ea a quibus natura vincimur' (*Aristotelis Quaestiones mechanicae*, tr. Niccolò Leonico Tomeo, in *Opuscula nuper in lucem aedita* (Venice, 1525), f. xxiii); see pseudo-Aristotle, *Mechanical Problems*, 847a11–29, ed. and tr. W.S. Hett, in *Aristotle, Minor Works* (Cambridge, Mass.: Harvard Univ. Press, 1963), 330–1. Cf. Moletti, *Discorso*, f. 155v (quoted in n. 4 above).

8 Note that for Moletti and his contemporaries, what we would call 'mechanical advantage' includes the ability of small military forces to withstand or defeat much larger forces with the aid of fortifications and other military machines. Cf. Filippo Pigafetta's similar statement in the Dedicatory Letter to his translation of Guido Ubaldo, *Liber Mechanicorum* (Pesaro, 1577), tr. Filippo Pigafetta, *Le mechaniche* (Venice, 1581), tr. in Drake and Drabkin, *Mechanics in Sixteenth-Century Italy*, 249.

9 'Huiusmodi autem sunt in quibus et minora superant maiora et quaecunque momentum parvum habent magna movent pondera. Et omnia fere illa que

Mechanica nuncupamus problemata' (*Aristotelis Quaestiones mechanicae,* tr. Leonico, f. xxiii); see pseudo-Aristotle, *Mechanical Problems,* 847a11–29, ed. and tr. Hett, 330–1.

10 Pliny, *Historia naturalis,* 7.37, 125, ed. and tr. H. Rackham, 10 vols. (Loeb Classical Library) (London: Heinemann; Cambridge, Mass.: Harvard Univ. Press, 1942), 2:588–91.

11 Plutarch, *Vita Marcelli,* xix. 4–6, in *Plutarch's Lives,* ed. and tr. Bernadotte Perrin, 11 vols. (Loeb Classical Library) (London: Heinemann; New York: Putnam's Sons, 1917), 5:486–7; cf. Moletti, *Expositio,* f. 165v; and see W.R. Laird, 'Archimedes among the Humanists,' *Isis* 82 (1991): 629–38, esp. 636–7.

12 Cf. 'mechanica scientia est quae nos docet superare ingentes difficultates aut artis aut naturae parva vi' (Moletti, *Expositio,* f. 160r).

13 Greek *βαναυσία,* Latin *ars sellularia;* cf. 'Sub factiva [*scil.* pars philosophiae], quae utile magis opus respicit quam honestum, ordinantur illae artes omnes, quas Graeci *βαναυσας* [*sic*], sive *βαναυσικὰς* dicunt, nos sellularias nuncupare possumus' (Alessandro Piccolomini, *In Mechanicas quaestiones Aristotelis paraphrasis* [Venice, 1565], f. 3v); 'Mechanicam autem scientiam voco, ex qua causae et principia, ad quamplurimas artes sellularias, exhauriri possunt, quae quidem artes vulgo non recte, mechanicae nuncupantur, nam potius sellulariae, sive Banausicae, ac humiles dici deberent' (f. 5r); and 'Quo factum est, ut omnes artes quae humiles et sordidae atque ideo illiberales sunt, abusione quadam, mechanicas vocaverimus, cum potius sellulariae seu banausicae dici debeant' (f. 7v).

14 Cf. 'Non enim tignarium adducam fabrum, quem tu summis caeterarum disciplinarum viris compares: fabri enim manus architecto pro instrumento est,' Leon Battista Alberti, *L'Architettura* [*De re aedificatoria*], ed. and tr. Giovanni Orlandi, 2 vols. (Milan: Edizioni il Polifilo, 1966), 1:2; see Pamela O. Long, 'The Contribution of Architectural Writers to a "Scientific" Outlook in the Fifteenth and Sixteenth Centuries,' *Journal of Medieval and Renaissance Studies* 15 (1985): 265–98, at 272; and cf. 'Est igitur ars mechanica ea quae ingenio cogitatione et machinatione exercetur, et non consistit in factione, sed in cogitatione et machinatione, et quae mechanicus machinatus est praecipit postmodum artificibus ut ab eis construantur. Et differt haec ars in hoc ab illis quae vulgo mechanicae appellantur, nam illae consistunt potius in factione' (Moletti, *Expositio,* f. 160r).

15 Cf. 'Da questa [i.e., meccanica] nasce l' arte del fortificare et del defendere, et offendere le città; poiché una fortezza si può con ragione dire machina' (Moletti, *Discorso,* f. 156v); and 'Ma passiamo hora all' arte del fortificare, parte della mecanica, et non dell' architettura, si come Vitruvio e molti altri vogliono' (f. 167r); Moletti goes on to consider Vitruvius's definition of archi-

tecture and to distinguish the art of fortification from it (ff. 167r–v). One of the many works left unfinished by Moletti is a treatise on fortification begun on 17 September 1575 (Milan, Biblioteca Ambrosiana MS. S 100 sup., ff. 252r–254r). On the relation between military architecture and mechanics in the Renaissance, see Catherine Wilkinson, 'Renaissance Treatises on Military Architecture and the Science of Mechanics,' in *Les traités d'architecture de la renaissance*, ed. Jean Guillaume (Paris: Picard, 1988), 467–76.

16 The question of whether mechanics is an art or a science is the topic of one section of Moletti's *Expositio*, ff. 195r–v, 199r; revised version, ff. 168r–v.

17 Aristotle, *Nicomachean Ethics*, 6.4, 1040a7–11.

18 See Moletti's discussion of subalternation in his *Discorso*, ff. 141v–142r; and in his *Expositio*, f. 165r; on the history of mechanics as a subalternated science, see W.R. Laird, 'The Scope of Renaissance Mechanics,' *Osiris* 2nd series, 2 (1986): 43–68; on the subalternation of sciences in general, see Steven J. Livesey, '*Metabasis*: The Interrelationship of the Sciences in Antiquity and the Middle Ages,' PhD diss. Univ. California, Los Angeles, 1982; and W.R. Laird, 'The *Scientiae Mediae* in Medieval Commentaries on Aristotle's *Posterior Analytics*,' PhD diss. Univ. Toronto, 1983.

19 'Sunt autem haec neque naturalibus omnino quaestionibus eadem neque seiugata valde: verum mathematicarum contemplationum naturaliumque communia' (*Aristotelis Quaestiones mechanicae*, tr. Leonicus, f. xxiii).

20 Cf. 'Mechanica enim ars, illa tantum existimanda est, qua excogitantes, quamplures machinas constructionesque caeteris artibus sellulariis invenimus. Quam philosophiae contemplatricis partibus annumerandum esse nemini dubium esse debet, quippe quae geometriae subiecta multarum artium principia excogitat ac speculatur. Quae quid est principia et si dirigantur ad opus, non ob id tamen horum principiorum facultas contemplatrix non existimabitur. Veluti et geometria et perspectiva, licet principia tradant pictori, qui opus respicit, contemplatrices tamen habentur et sunt. Scientia igitur magis quam ars facultas haec mechanica merito nuncupabitur, praesertim cum Aristoteles ipse, non solum ab initio *Mechanicarum questionum* sed etiam in <*De*> *generatione animalium*, in *Metaphysicis*, et quam pluribus aliis in locis de scientiis loquens, nomine artis abutatur. Haec itaque vel scientia vel arte quam mechanicam dicimus, his excellimus in quibus a natura vinceremur. Ea autem sunt, tum quibus maiora a minoribus superantur, tum etiam quae et si modicae virtutis sint, graviora tamen commovent, tum demum ea omnia fere, quae mechanicis quaestionibus inquiruntur. Quae quidem, quamvis non penitus naturales, nec omnino mathematicae dici possint, sed utrorumque naturam aptant, magis tamen mathematicis quaestionibus propinquae sunt. Cum enim in naturali materia versentur haec, v<e>luti circa mobilia et pon-

derosa, ea ratione qua huiusmodi sunt (circa enim lapides, ligna, et alia id genus, subiectae mechanico versantur artes) modo autem mathematico per designationes et proportiones demonstrentur. Ac quaelibet facultas magis a modo ostendendi quam a subiecta materia nuncupari debeat: iccirco haec facultas, potius mathematicis contemplationibus, quam naturalibus annumerabitur' (Piccolomini, *In Mechanicas quaestiones Aristotelis paraphrasis*, ff. 8r–v).

21 Proclus, *Procli Diadochi in primum Euclidis Elementorum librum commentarii*, ed. Godfried Friedlein (Leipzig, 1873), 38–41; tr. Glenn R. Morrow, *A Commentary on the First Book of Euclid's Elements*, (Princeton: Princeton Univ. Press, 1970; 2nd ed. 1992), 31–4. Moletti probably used the translation by Francesco Barozzi: Proclus, *Procli Diadochi Lycii in primum Euclidis Elementorum commentariorum libri iv* tr. Francesco Barozzi (Padua, 1560); on Barozzi, see Paul Lawrence Rose, 'A Venetian Patron and Mathematician of the Sixteenth Century: Francesco Barozzi (1537–1604),' *Studi Veneziani*, new series 1 (1977): 119–78.

22 In what was probably the original draft of the *Dialogue*, Moletti went off here on an extensive digression concerning the status of mathematics and its relation to other sciences, a digression that is preserved, in Moletti's own hand, in Milan, Biblioteca Ambrosiana, MS. D 235 inf., ff. 59r–62v (= D), which is edited and translated in the appendix below.

23 The subject, principles, and properties of mechanics are the topic of another section of Moletti's *Expositio*, ff. 165r–167v.

24 Livy, *Ab urbe condita*, XXIV, 34; ed. C.F. Walters and R.S. Conway, 5 vols. (Oxford: Clarendon Press, 1914), vol. 3; this passage is also cited by Moletti in his *Expositio*, f. 159v; Plutarch, *Vita Marcelli*, xiv. 7–8, in *Plutarch's Lives*, ed. and tr. Perrin, 5:472–3).

25 Proclus, *In primum Euclidis Elementorum*, ed. Friedlein, 63; tr. Morrow, 51; cf. 'Ha per soggetto quest' arte ò questa scienza principale secondo il parer mio la machina, ò quella cosa che può operare con maraviglia, come verbi gratia per venir all' essempio: onde è che con la forza di un sol huomo si fa quella operatione che non possono fare appena mille huomini, si come leggiamo haver fatto fare Achimede ad Hierone Re di Siracusa. Percioche havendo Hierone fatto fare una gran nave con intentione di mandarla à Tolomeo Re di Egitto; et affaticandosi molti anzi tutti gli' huomini della città per mandarla in acqua, et in somma in vano. Dopo che Archimede hebbe veduto ciò disse ad Hierone che voleva che egli solo sedendo nella sua sedia la varasse et la mandasse in acqua. Stupefatto il Re di ciò senza dare molta fede alla parole d' Archimedi, aspettò il tempo, il quale venuto havendo Archimede appoggiatta la machina alla nave diede una corda in mano al Re e disseli, che la tirasse alquanto; et subito senza che egli si movesse da sedere vidde movere

la nave, e fra poco la vidde poi tutta in acqua. Di che maravigliato disse che tutto quello che da Archimede da indi inanzi le fosse detto crederebbe' (Moletti, *Discorso*, ff. 155v–156r).

26 'La machina è una perpetua e continuata congiuntione di materia, che ha grandissima forza, ai movimenti dei pesi' (Vitruvius, *I dieci libri dell'Architettura di Vitruvio*, 10.1.1, tr. Danielo Barbaro (Venice, 1556), 254; cf. Vitruvius, *De architectura libri decem*, 10.1.1; ed. and tr. Louis Callebat and Philippe Fleury, in *Vitruve, De l'Architecture livre X* (Paris: Belles Lettres, 1986), 4; and *Vitruvius on Architecture*, ed. and tr. Granger, 2:274–5.

27 'Di queste trattorie altre si movono con machine, altre con istrumenti, et pare, che tra machina e strumento ci sia questa differenza, che bisogna che le machine con più opere, overo con forza maggiore conseguano gli effetti loro, come le baliste et i preli dei torcolari. Ma gli istrumenti col prudente toccamento d' un' opera fanno quello, che s' hanno proposto di fare come sono gli involgimenti degli scorpioni, et degli circoli diseguali' (Vitruvius, *I dieci libri dell'Architettura di Vitruvio*, 10.1.3, tr. Barbaro, 255); cf. *Vitruve, De l'Architecture livre X*, ed. and tr. Callebat and Fleury, 5; and *Vitruvius on Architecture*, ed. and tr. Granger, 2:276–7.

28 Cf. 'Hora al proposito noi veggiamo qui l' effetto mirabile, che la sola forza della machina haverà varato la nave. È adunque da vedere per qual cagione o con qual virtù questa machina l' haverà fatto et questo è considerare le proprietà della machina. Adonque se della machina se considera la proprietà, la machina viene ad essere soggetto. Ma è da avvertire, che io non solo per machina intendo come sarebbe à dire l' argana ò le taglie ò la statera ò altro tale instrumento, ma ancora tutto quello che può esteriormente et che huomo non sia aiutare la forza dell' huomo o accrescere quella, si come è la stanga, ò cosa tale. Ben vero è che Vitruvio fa differenza tra machina et organo o istromento, volendo che machina sia quella dove sia molto lavoro, et organo dove poco, ò semplice instromento sia si come egli da l' essempio del torcolo che preme l' uva ò l' oglio, ò cosa tale, et lo scorpione, o come tra l' horivolo da ruote et la stanga, ò cosa tale. Nondimeno io però non voglio farvi differenza alcuna, ma co 'l nome di machina chiamo tutti gli instromenti che aiutano ò possano aiutare la forza dell' huomo; non nego che non vi sia differenza della quale altrove' (Moletti, *Discorso*, f. 156r).

29 Cf. Vitruvius, *I dieci libri dell'Architettura di Vitruvio*, 10.1.4, tr. Barbaro, 255; cf. *Vitruve, De l'Architecture livre X*, ed. and tr. Callebat and Fleury, 5; and *Vitruvius on Architecture*, ed. and tr. Granger, 2:276–7.

30 Vitruvius, *I dieci libri dell'Architettura di Vitruvio*, 10.1.4, tr. Barbaro, 255; cf. *Vitruve, De l'Architecture livre X*, ed. and tr. Callebat and Fleury, 5; and *Vitruvius on Architecture*, ed. and tr. Granger, 2:276.

31 'Omnium autem huiusmodi causae principium habet circulus: istud vero ratione contingit. Ex admirabili etenim mirandum accidere quippiam non est absurdum' (*Aristotelis Quaestiones mechanicae*, tr. Leonico, f. xxiiii recto); see pseudo-Aristotle, *Mechanical Problems*, 847 b 16–19, ed. and tr. Hett, 332–3. Cf. Moletti, *Discorso*, f. 157r.

32 Plutarch, *Vita Marcelli*, xiv. 7–8; in *Plutarch's Lives*, ed. and tr. Perrin, 5:473; cf. Pappus Alexandrinus, *Collectio mathematica*, 8, prop. 11, tr. Federico Commandino (Pesaro, 1588); in his lectures on the *Mechanical Problems* Moletti cites book 8 of the *Collectio mathematica* (*Expositio*, f. 183r), and a little later notes that Commandino's translation has not yet been published: 'Post hos scriptores qui di machinationibus scripserunt, inveniuntur *Collectoria* Pappi Alexandrinii, a clarissimo Federico Commandino latina facta, sed non dum tipis data' (f. 184r). A lever is not explicitly mentioned by either Plutarch or Pappus. In his *Discorso*, Moletti quotes the phrase as 'Dammi un luogo dove possa appoggiare la machina, et io levarò la terra dal centro et la mandarò ad un altro luogo' (f. 163r).

33 'Maxime autem est admirandum simul contraria fieri. Circulus vero ex huiusmodi est constitutus, statim enim ex commoto effectus est et manente, quorum natura ad se invicem est contraria. Quamobrem isthaec cernentes minus admirari convenit contingentes in illo contrarietates. In primis enim lineae illi quae circuli orbem amplectitur nullam habenti latitudinem contraria quodammodo inesse apparent, concavum scilicet et curvum. Haec autem eo a se invicem distant modo quo magnum et parvum; illorum et enim medium est aequale, horum vero rectum. Quapropter cum ad se invicem commutantur illa quidem prius aequalia fieri necesse est quam extremorum utrunlibet, lineam vero rectam quando ex curva concava, aut ex huiusmodi rursum curva fit et circularis. Unum quidem igitur istuc absurdum inest circulo; alterum autem quod simul contrariis movetur motionibus: simul enim ad anteriorem movetur locum et ad posteriorem, et ea quae circulum describit linea eodem se habet modo. Ex quo enim incipit loco illius extremum ad eundem rursus redit; illa enim continuo commota extremum rursus efficitur primum; quamobrem manifestum quod inde mutatum est. Quapropter (ut dictum est prius) non est inconveniens ipsum miraculorum omnium esse principium. Ea igitur quae circa libram fiunt ad circulum referuntur; quae vero circa vectem ad ipsam libram. Alia autem fere omnia quae circa mechanicas sunt motiones ad vectem' (*Aristotelis Quaestiones mechanicae*, tr. Leonico, f. xxiiii recto [see also Leonico's commentary, ff. xxiiii verso–xxv recto]); see pseudo-Aristotle, *Mechanical Problems*, 847b16–848a15, ed. and tr. Hett, 332–5; and cf. Piccolomini, *In Mechanicas quaestiones Aristotelis paraphrasis*, ff. 9r–v.

34 'Praeterea etiam quoniam unica existente quae ex centro est linea nullum aliud alii quae in illa sunt punctorum aequa velocitate feratur, sed citius semper quod a manente termino est remotius; plaeraque [= pleraque] miraculorum accidunt in circuli motionibus, de quibus in hiis quae posthac adducentur quaestionibus erit manifestum' (*Aristotelis Quaestiones mechanicae*, tr. Leonico, f. xxiiii recto [see also Leonico's commentary, f. xxv recto–verso]); see pseudo-Aristotle, *Mechanical Problems*, 848a15–19, ed. and tr. Hett, 334–5; cf. Piccolomini, *In Mechanicas quaestiones Aristotelis paraphrasis*, f. 9r; and Moletti: 'Ho detto tutte le cose mechanice poter nascere da due principii, l' uno essere il cerchio et l' altro essere lo spirito o 'l vacuo; che l' uno et l' altro si può dire come appresso dirò. Et però Aristotele parlando de problemi mechanici disse, che tutti si riducono alla stanga da lui detta vectis; questa si riduce alla libra, et ultimamente questa se [= si] riduce al cerchio, il quali è cagione di tutte le maraviglie di questa sorte. Nel cerchio dunque da se solo considerato s' ha da considerare questo, che quanto più le parti sue s' allontanano dal centro, con tanta maggior velocità se muovono, il che lo verro prima manuducendo a questo modo, e poi ne darò dimostratione e dichiarirò le parole d' Aristotele' (*Discorso* f. 157r).

35 'Cicius [*i.e.*, citius] enim bifariam dicitur: sive enim in minori tempore equalem pertransit locum citius fecisse dicimus, seu in equali maiorem. Maior autem in equali tempore maiorem describit circulum; qui enim extra est maior eo qui intus est' (*Aristotelis Quaestiones mechanicae*, tr. Leonico, ff. xxv verso–xxvi recto); see pseudo-Aristotle, *Mechanical Problems*, 848b6–9, ed. and tr. Hett, 336–7.

36 That the two movements that combine to describe a circle are devoid of any fixed ratio is the supposition proved as proposition V below. That two movements in any fixed ratio describe a straight line he calls the 'first supposition' and proves as proposition II below. In all this he is following and expanding upon the proofs in the *Mechanical Problems*: cf. *Aristotelis Quaestiones mechanicae*, tr. Leonico, ff. xxvi recto–xxvii recto; see pseudo-Aristotle, *Mechanical Problems*, 848b10–849a6, ed. and tr. Hett, 336–41. And cf. Moletti's briefer treatment of this in *Discorso*, ff. 159v–160r.

37 See *Aristotelis Quaestiones mechanicae*, tr. Leonico, ff. xxvii recto–verso; and pseudo-Aristotle, *Mechanical Problems*, 849a6–b19, ed. and tr. Hett, 340–7. Cf. Moletti's discussion of this, including a diagram different from figure I but identical to Leonico's, in his *Discorso*, ff. 160r–161r.

38 'The straight lines joining equal and parallel straight lines (at the extremities which are) in the same directions (respectively) are themselves also equal and parallel,' Euclid, *Elements*, 1.33, tr. Sir Thomas L. Heath, *The Thirteen Books of the Elements*, 3 vols, 2nd ed. (1926; rpt. New York: Dover, 1956),

1:322–3; 'In parallelogrammic areas the opposite sides and angles are equal to one another, and the diameter bisects the areas,' Euclid, *Elements*, 1.34, tr. Heath, 1: 323–6; 'In a circle equal straight lines are equally distant from the centre, and those which are equally distant from the centre are equal to one another,' Euclid, *Elements*, 3.14, tr. Heath, 2:34–6.

39 On the two suppositions, see note 36 above.

40 'If from a parallelogram there be taken away a parallelogram similar and similarly situated to the whole and having a common angle with it, it is about the same diameter with the whole,' Euclid, *Elements*, 6.26, tr. Heath, 2:255–6.

41 'Horum autem causa quoniam duas fertur lationes ea quae circulum describit linea. Quando quidem igitur in proportione fertur aliqua id quod fertur super rectam ferri necesse est, et haec diameter efficitur figurae quam faciunt illae quae in huiusmodi proportione coaptantur lineae. Sit enim proportio secundum quam latam fertur quam habet AB ad AC. Et A quidem feratur versus B, AB vero subterferatur versus MC. Latum autem fit A quidem ad D. Ubi autem est AB versus E. Quantum igitur lationis erat proportio quam AB habet ad AC, necesse est et AD ad AE hanc habere proportionem. Simile igitur est proportione parvum quadrilaterum maiori, quemobrem et eadem illorum est diameter, et A erit ad F. Eodem etiam ostendetur modo ubicunque latio deprendatur: semper enim supra diametrum erit. Manifestum igitur quod id quod secundum diametrum duabus fertur lationibus necessario secundum laterum proportionem fertur. Si enim secundum aliam quampiam, non feretur secundum diametrum' (*Aristotelis Quaestiones mechanicae*, tr. Leonico, f. xxvi recto; and pseudo-Aristotle, *Mechanical Problems*, 848b11–28, ed. and tr. Hett, 336–9. Cf. 'Ma vediamo come ciò dice Aristotele. Egli adunque nel quarto miracolo del cerchio dice queste parole: «In primis igitur quae accidunt circa libram, etc.», et seguendo «huius autem <rei> principium est quamobrem in ipso circulo, quae plus a centro distat linea, eadem <vi> commota citius fertur quam illa quae minus distat» et segue; et appresso poi dice: «horum causa quoniam duas fert lationes ea quae circulum describit linea». Et in summa, per non portare tutte le parole d' Aristotele, dice ciò nascere perché la linea che descrive il cerchio si move con dui movimenti, i quali sono disgionti d' ogni proportione in ogni tempo et sono contrarii; perché l' uno è naturale, et l' altro è contra natura, perché quando congionti in alcuna proportione fossero, descriverebbono una linea retta, si come egli dimostra chiaramente' (Moletti, *Discorso*, ff. 159v–160r); except for the two omissions that I have restored, Moletti's Latin quotations here are word for word from Leonico's translation.

42 This would be one of the instruments mentioned at the beginning of the

Dialogue, which mechanically produces a diagonal straight line from two combined movements. Moletti himself possessed certain mathematical instruments: in Milan, Biblioteca Ambrosiana MS. S 94 sup., f. 171, are several lists of instruments, one labelled as those belonging to Moletti. And in his Testament of 1587 Moletti mentions his mathematical instruments, otherwise unspecified, though he leaves to Pinelli, along with some books, a bronze artillery piece (Milan, Biblioteca Ambrosiana MS. S 98 sup., ff. 184–9; printed in Favaro, *Amici et corrispondenti di Galileo Galilei*, 113–18; no instruments are mentioned in his earlier Testament of 1570 (Venice, Archivio di Stato, Sezione Notarile: Testamenti, busta 646, Notaio Michieli Francesco, Testamento no. 441; printed in Favaro, *Amici et corrispondenti*, 102–4).

43 'In any triangle, if one of the sides be produced, the exterior angle is equal to the two interior and opposite angles, and the three interior angles of the triangle are equal to two right angles' (Euclid, *Elements*, 1.32, tr. Heath, 1:316–22).

44 'If in a triangle two angles be equal to one another, the sides which subtend the equal angles will also be equal to one another' (Euclid, *Elements*, 1.6, tr. Heath, 1:255–8).

45 This quip alludes to the Aristotelian doctrine that circular and straight lines – and consequently the motions along them – are utterly distinct from each other: see Aristotle, *Physics*, 5.4, 227b14–20.

46 Nicholas Copernicus, *De revolutionibus orbium coelestium*, 3.4, ed. Jerzy Dobrzychi, tr. Edward Rosen, *Nicholas Copernicus Complete Works*, 3 vols. (Warsaw/Cracow: Polish Academy of Sciences, 1972–85), 2:125–6.

47 Johannes de Sacrobosco, *Tractatus de sphaera*, ed. and tr. Lynn Thorndike, *The Sphere of Sacrobosco and Its Commentators* (Chicago: Univ. Chicago Press, 1949); Sacrobosco's *Sphere*, the standard introduction to astronomy since its composition in the mid-thirteenth century, was still taught as an elementary text in Moletti's time. By 'Euclid' here Moletti probably means the *Elements*, another standard text at the time.

48 On Moletti's knowledge and opinion of Copernicus, see the Introduction, 27.

49 PR is referring to having heard lectures on the pseudo-Aristotelian *Mechanical Problems*, perhaps at a university. Pietro Catena, Moletti's immediate predecessor in the chair of mathematics at the University of Padua, is the only lecturer before Moletti himself known to have lectured at a university on this text. Guidobaldo heard these lectures in 1564, as did Bernardino Baldi in 1573 (see the Introduction, 17–18), but whether Moletti himself heard them, or whether he heard the text expounded by his teacher in Messina, Francesco Maurolico, is unknown. Perhaps the reference is simply to establish the

Prince's authority in mechanics. Like the previous proposition, this one has no precedent in the *Mechanical Problems*, and perhaps is, as Moletti claims, original with him.

50 On this proposition, see the Introduction, 27–8 above.

51 'If a straight line falling on two straight lines make the alternate angles equal to one another, the straight lines will be parallel to one another' (Euclid, *Elements*, 1.27, tr. Heath, 1:307–9); 'The straight lines joining equal and parallel straight lines (at the extremities which are) in the same directions (respectively) are themselves also equal and parallel' (Euclid, *Elements*, 1.33, tr. Heath, 1:322–3).

52 'Straight lines parallel to the same straight line are also parallel to one another' (Euclid, *Elements*, 1.30, tr. Heath, 1:314–15).

53 That is, by Euclid, *Elements*, 3.28 (see the next note).

54 'In equal circles equal straight lines cut off equal circumferences, the greater equal to the greater and the less to the less' (Euclid, *Elements*, 3.28, tr. Heath, 2:59–60); 'If in a circle a straight line through the centre bisect a straight line not through the centre, it also cuts it at right angles; and if it cut it a right angles, it also bisects it' (Euclid, *Elements*, 3.3, tr. Heath, 2:10).

55 'Si autem in nulla feratur proportione secundum duas lationes, nullo in tempore rectam esse lationem est impossibile. Sit enim recta. Posita igitur hac pro diametro et circumrepletis lateribus, illud quod fertur secundum laterum proportionem ferri necesse est: hoc enim demonstratum est prius. Non igitur rectam efficiet id quod secundum nullam proportionem in nullo fertur tempore. Si autem secundum quampiam feratur proportionem et in tempore quopiam, hoc necesse est tempus rectam esse lationem, per ea quae retro sunt dicta. Quamobrem circulare est id quod secundum nullam proportionem nullo in tempore duas fertur lationes' (*Aristotelis Quaestiones mechanicae*, tr. Leonico, ff. xxvi recto–xxvi verso); and pseudo-Aristotle, *Mechanical Problems*, 848b28–35, ed. and tr. Hett, 338–41; cf. Moletti, *Discorso*, ff. 160r–162r.

56 Cf. 'It [i.e., the weight] is heavier in descending, to the degree that its movement toward the centre (of the world) is more direct,' Jordanus, *De ratione ponderis*, ed. and tr. E.A. Moody and Marchall Clagett, in *The Medieval Science of Weights* (Madison: Univ. Wisconsin Press, 1952), 175; and 'Also we request that it be conceded that a heavy body in descending is so much the heavier as the motion it makes is straighter toward the center of the world' (Niccolò Tartaglia, *Quesiti*, book 8, qu. 24, 3rd petition; tr. in Drake and Drabkin, *The Science of Mechanics in Sixteenth-Century Italy*, 118, where a diagram similar to Moletti's is found). Cf. Moletti's similar treatment in the *Discorso*, f. 157r–v.

57 Cf. Tartaglia, *Quesiti*, book 8, qu. 24, 3rd petition; tr. in Drake and Drabkin,

The Science of Mechanics in Sixteenth-Century Italy, 119, where again a similar diagram is found; and Moletti, *Discorso*, f. 158r–v.

58 I do not know whom Moletti means here.

59 The angle of contingence or 'horn' angle is the angle between an arc and its tangent. Euclid, in *Elements* 3. 16, proved that all horn angles, whatever the size of the circle, are less than any rectilinear angle (tr. Heath, 2:37–9). In the *Discorso* (ff. 158v–159r) Moletti argued that here Euclid meant only that horn angles cannot be compared to rectilinear angles, and not that they are not quantities and so cannot even be compared to one another, rebutting the arguments of Jacques Peletier in his edition of the *Elements* of 1557. For a brief account of the controversy over the horn angle and of Peletier's arguments, see Heath, *Euclid's Elements*, 2:39–43.

60 'Out of three straight lines, which are equal to three given straight lines, to construct a triangle: thus it is necessary that two of the straight lines taken together in any manner should be greater than the remaining one' (Euclid, *Elements*, 1.22, tr. Heath, 1:292–4).

61 'If on one of the sides of a triangle, from its extremities, there be constructed two straight lines meeting within the triangle, the straight lines so constructed will be less than the remaining two sides of the triangle, but will contain a greater angle' (Euclid, *Elements*, 1.21, tr. Heath, 1:289–92).

62 'If in a circle a straight line through the centre bisect a straight line not through the centre, it also cuts it at right angles; and if it cut it at right angles, it also bisects it' (Euclid, *Elements*, 3.3, tr. Heath, 2:10–11). It is unnecessary, however, to prove angles ELB and CHA to be right angles, since lines IAK and MBN were drawn perpendicular to lines CD and EF in the first place.

63 'In a circle the angle in the semicircle is right, that in a greater segment less than a right angle, and that in a less segment greater than a right angle; and further the angle of the greater segment is greater than a right angle, and the angle of the less segment less than a right angle' (Euclid, *Elements*, 3.31, tr. Heath, 2:61–5).

64 'In any triangle, if one of the sides be produced, the exterior angle is equal to the two interior and opposite angles, and the three interior angles of the triangle are equal to two right angles' (Euclid, *Elements*, 1.32, tr. Heath, 1:316–22).

65 'If two triangles have the two sides equal to two sides respectively, and have the angles contained by the equal straight lines equal, they will also have the base equal to the base, the triangle will be equal to the triangle, and the remaining angles will be equal to the remaining angles respectively, namely those which the equal sides subtend' (Euclid, *Elements*, 1.4, tr. Heath, 1:247–50).

66 'On a given straight line and at a point on it to construct a rectilineal angle equal to a given rectilineal angle' (Euclid, *Elements*, 1.23, tr. Heath, 1:294–6); 'If two triangles have the two sides equal to two sides respectively, and have also the base equal to the base, they will also have the angles equal which are contained by the equal straight lines' (Euclid, *Elements*, 1.8, tr. Heath, 1:261–4).

67 'About a given triangle to circumscribe a circle' (Euclid, *Elements*, 4.5, tr. Heath, 2:88–90).

68 'Similar segments of circles on equal straight lines are equal to one another' (Euclid, *Elements*, 3.24, tr. Heath, 2:53–4).

69 'In any triangle, if one of the sides be produced, the exterior angle is equal to the two interior and opposite angles, and the three interior angles of the triangle are equal to two right angles' (Euclid, *Elements*, 1.32, tr. Heath, 1:316–17); thus the segments EOF and CID are similar – by *Elements*, 3. definition 11, 'Similar segments of circles are those which admit equal angles, or in which the angles are equal to one another' (tr. Heath, 2:2) – which is required for 3.24.

70 Cf. 'The circumference of a smaller circle is more curved; that of a larger circle is less curved' (Blasius of Parma, *Tractatus de ponderibus*, pars prima, prop. IV; ed. and tr. Moody and Clagett, in *The Medieval Science of Weights*, 240–1).

71 In this awkward sentence Moletti seems to be referring to proving that from equal chords in unequal circles, the versed sine ('the rest or residual of the diameter') in the larger circle will be less than the versed sine in the smaller: i.e., in figure IX, that IH is less than ML and so arc CID is 'closer' to line CD than arc EMF is to line EF.

72 By specifying that CG be larger than CB, Moletti has committed a *petitio principii*: the demonstration is to prove that CG is larger than CB.

73 'In equal circles equal angles stand on equal circumferences, whether they stand at the centres or at the circumferences' (Euclid, *Elements*, 3.26; tr. Heath, 2:58–9); Moletti (or his copyist) had mistakenly cited 3.27, which is the converse. Note that 3.26 applies only to equal arcs and equal angles in equal circles, and not to larger or smaller ones, as Moletti wants it to.

74 'Similar segments of circles on equal straight lines are equal to one another' (Euclid, *Elements*, 3.24; tr. Heath, 2:53–4). Again Moletti implicitly extends this to larger and smaller lines.

75 Moletti touches on this in the Second Day: see below, 149.

76 Proclus, *In primum Euclidis Elementorum librum commentarii*, ed. Friedlein, 38–9; tr. Morrow, 31–2.

77 On the relation between mechanics and the military arts, including fortification and deployment of forces, see Moletti, *Discorso*, ff. 169r–170v.

78 On the division of mechanics see Moletti, *Expositio*, ff. 170r–174v; see especially the chart of the divisions of mechanics on f. 171r, and the huge 'Partitio mathematicarum scientiarum Josephi Moleti,' Milan, Biblioteca Ambrosiana MS. R 94 sup., f. 184. Cf. Moletti, *Discorso:* 'Le parti della mechanica sono due, poiché dui sono i suoi principii, l' uno è il cerchio et il secondo è lo spirito ò l' aria ò 'l vacuo. Tutti gli altri principii si riducono a questi dui, perché tutte le cose de gli instromenti ò machine belliche de gli antichi, come l' ariete e gli altri tutti, si riducono al cerchio, si come io dimostrarò altrove, e tutte le machine d' altra sorte, come l' artigliaria ò altre tali, si riducono al vacuo ò allo spirito. Possiamo poi comporre noi questi dui principii, ò fare alcuni instromenti c' habbino l' uno e l' altro principio in se, come sono le trombe delle navi, e quelle machine con le quali si tira l' acqua dal fondo de pozzi nel piano. Ne io mi credo, che si possa trovar machina, che facci forza che non si reduca a questi dui principii, ò all' uno solo, ò all' altro, ò à tutti due insieme' (*Discorso*, f. 156r–v).

79 This is Archimedes' treatise on the elements of mechanics, where he proves the law of the lever and develops the idea of centres of gravity. The *De Aequeponderibus* (as it was known in Latin), together with *De mensura circuli, De quadratura parabolae*, and book 1 of *On Floating Bodies*, was first printed by Tartaglia, in Moerbeke's Latin translation, in 1543; for Tartaglia, in contrast to Moletti, the causes and principles of mechanics are found in the science of weights: *Quesiti* (Venice, 1554), book 7, f. 78r; cf. Moletti in the *Discorso:* 'Sotto quella [i.e., meccanica] si comprende la scienza de pesi, e quel libro d' Archimede, *De aequiponderantibus et de centris gravitatum*; et questa fu quella della quale Archimede fu padrone, et la quale lo fece stimare miracolo del mondo' (*Discorso*, f. 156r). Moletti apparently never wrote further on the science of weights.

The Second Day

1 The Palazzo Te was the Gonzagas' palatial retreat on the outskirts of Mantua.

2 Pseudo-Aristotle, *Quaestiones mechanicae*, qu. 34, 858a23–b3; tr. Leonico, f. liii verso; on the sling, which works on the same principle as AN's spear-thrower here, see qu. 12, 852a38–b10; tr. Leonico, f. xxxvii verso; and Cardano, *Opus novum de proportionibus*, prop. 113 (Basle, 1570; rpt. in *Opera omnia*, London, 1662–3), 4:517.

3 See below, 141–7.

4 This is question 3 of the *Mechanical Problems*, 850a30–b10; tr. Leonico, ff. xxxi verso–xxxii recto.

5 Plutarch, *Lives*, xiv, 7; tr. Perrin, 473; cf. Moletti, *Expositio*, f. 178v.

6 Aristotle, *De motu animalium*, 698a1–b8; ed. and tr. E.S. Forster (London: Heinemann; Cambridge, Mass.: Harvard Univ. Press, 1959); and Niccolò Leonico Tomeo, *Paraphrasis in libellum de animalium motu Aristotelis*, printed with his translation of the *Quaestiones mechanicae* in his *Opuscula nuper in lucem aedita* (Venice, 1525), ff. xv recto–verso; cf. Moletti's similar discussion in the *Discorso:* 'Tanto voglio solo avvertire, che volendo studiare tale scienza, fa di mestiere, che molto bene s' apprendano i principii sudetti. Et appresso avvertire, come nelle questioni s' applicano. Come se si dicesse et si ricercasse la cagione apunto della stanga, per non passare per ora ad altre questioni, et si dicesse: Onde è che volendo noi movere un gran peso da terra senza alzarlo da quella, quello dalla forza di molti huomini non può essere mosso, i quali con le mani sole senza altro instromento si sforzino di moverlo; et nondimeno applicata vi la stanga da minor numero di huomimi si muove facilmente. Volendo solvere la dimanda, noi sappiamo, che dovemo ricorrere al principio messo et dire: la stanga si riduce alla libra o bilancia, et questa al cerchio. Ma come la ridurremo al cerchio? Faremo cosi: sappiamo noi prima, che la stanga non move se non è ad alcuna cosa stabile appoggiata; però è di bisogno vedere dove si appoggia. Ch' abbia ad appoggiarsi ad alcuna cosa stabile è manifesto per quello che prova Aristotele nel libretto *De motibus animalium*, dove dimostra che ogni cosa che si move, conviene che si muova sopra ad alcuna stabile; come dovendosi movere il piede, conviene che la gamba stia ferma, et dovendosi movere la coscia, conviene che 'l resto del corpo stia fermo, et così del resto. Così ancora se l' huomo si muove su l' arena, fa meno viaggio che se si muove su 'l terreno saldo; il che nasce da questo, che si muove su una cosa non del tutto stabile. A questo avvertendo Archimede disse al suo Re: Dammi un luogo dove possa appoggiare la machina, et io levarò la terra dal centro et la mandarò ad un altro luogo' (*Discorso*, ff. 162v–163r). Moletti goes on to illustrate the principle of the lever and to criticize Tartaglia's treatment of question 1 of the *Mechanical Problems* (ff. 163r–165v).

7 Aristotle, *De motu animalium*, 2, 698b8–18; Leonico, ff. xv verso–xvi recto.

8 Aristotle, *De incessu animalium*, 709a25–b4; ed. and tr. E.S. Forster (London: Heinemann; Cambridge, Mass.: Harvard Univ. Press, 1959); and Niccolò Leonico Tomeo, *De animalium progressu paraphrasis liber*, also printed with his translation of the *Quaestiones mechanicae* in *Opuscula nuper in lucem aedita* (Venice, 1525); Leonico discusses the motion of snakes, with a marginal diagram similar to Moletti's, on f. vii recto. This subject does not come up again in the *Dialogue*.

9 Aristotle, *De motu animalium*, 698b21–699a11; Leonico, *Paraphrasis de animalium motu*, ff. xv verso–xvi recto.

10 See Aristotle, *De motu animalium*, 698b15; Leonico, *Paraphrasis de animalium motu*, f. xv verso.

11 Athough these topics are not raised again in the *Dialogue*, the question of whether nature in general makes use of mechanics is the subject of an extensive section of Moletti's *Expositio*, ff. 175r–179r.

12 ὑπομόχλιον, Aristotle, *Quaestiones mechanicae*, qu. 3, 850a36; Leonico uses the Greek word transliterated into Latin (f. xxxii recto); see Vitruvius, *De architectura*, 10.3.2; tr. Barbaro, 10. cap. 8, 259.

13 In Aristotelian physics, only the first mover (*primum movens*), which is incorporeal, is itself unmoved: see Aristotle, *Physics*, 8.10, 267b9–26. For a general introduction to the medieval physics of motion, see John E. Murdoch and Edith D. Sylla, 'The Science of Motion,' in *Science in the Middle Ages*, ed. David C. Lindberg (Chicago: Univ. Chicago Press, 1978), 206–64.

14 Besides local motion (i.e., motion from place to place), the other species of motion or change in Aristotelian physics are qualitative change (change from one quality to another) and quantitative change (change in size); see Aristotle, *Physics*, 5.1, 225b5–9; see also 3.1, 200b33–34.

15 Aristotle, *Physics*, 3.1, 201a19–27; the only mover that is not changed or moved itself by moving something else is, as Aristotle suggests in this passage, the prime mover, which is not a *natural* agent. In late medieval Aristotelian physics the effect of moving something else on the mover was called *reactio*: see Marshall Clagett, *Giovanni di Marliano and Late Medieval Physics* (New York: Columbia Univ. Press, 1941), esp. 34–58; and Anneliese Maier, *Die Vorläufer Galileis* (Rome: Storia e Letteratura, 1949), 73–8.

16 This is standard medieval Aristotelian physics. Motion is either natural or unnatural (violent) depending on whether it expresses the intrinsic nature of the thing moved or is imposed from outside and against that nature (see Aristotle, *Physics*, 2.1, 192b8–13; and *De caelo*, 1.2, 268b27–269b16). See James A. Weisheipl, 'Natural and Compulsory Movement,' *New Scholasticism* 29 (1955): 50–81; rpt. in James A. Weisheipl, OP, *Nature and Motion in the Middle Ages*, ed. William E. Carroll (Washington: Catholic Univ. America Press, 1985), 25–48. Natural philosophy was concerned mainly with natural motions, while mechanics by definition was concerned with motions that often were against nature – see above 8, 12 and W.R. Laird, 'The Scope of Renaissance Mechanics,' *Osiris*, new series 2 (1986): 43–68.

17 Cf. 'Both heaviness, and lightness, of the thing moved, seem to thwart the force of the mover' (Jordanus de Nemore, *De ratione ponderis*, R4.10, ed. and tr. in Moody and Clagett, *The Medieval Science of Weights*, 218–19). In medieval ratio theory, ratios are 'denominated' by a name expressing the relation of their two terms; where the quotient of the terms is an integer or a rational

fraction, that quotient is the denomination: e.g., the ratio of 2 to 1 is denominated *dupla* (double), 1 to 2 is denominated *subdupla* (half), and 3 to 2 is denominated *sesquialtera*. For a ratio to be denominated by a very large number means (in modern parlance) that the ratio is very large; by 'imperceptible ratio' Moletti seems to mean the ratio of two terms whose quotient is vanishingly small. On medieval ratio theory see Michael S. Mahoney, 'Mathematics,' in *Science in the Middle Ages*, ed. Lindberg, 145–78, on 162–8; but see also John E. Murdoch, 'The Medieval Language of Proportions,' in *Scientific Change*, ed. A.C. Crombie (New York: Basic Books, 1963), 237–71. For ratio theory as applied to motion and change, see Clagett, *Giovanni Marliani*, 125–44; Anneliese Maier, *Die Vorläufer Galileis* (Rome: Storia e Letteratura, 1949), 81ff; tr. in part in Steven D. Sargent, *On the Threshold of Exact Science: Selected Writings of Anneliese Maier on Late Medieval Natural Philosophy* (Chicago: Univ. Chicago Press, 1982), 62–75; Marshall Clagett, *The Science of Mechanics in the Middle Ages*, 421–503; and Stillman Drake, 'Medieval Ratio Theory vs Compound Medicines in the Origins of Bradwardine's Rule,' *Isis* 64 (1973): 67–77.

18 I.e., a 'shield,' a silver coin worth one Venetian ducat.

19 Cf. 'Ex nulla proportione aequalitatis vel minoris inaequalitatis motoris ad motum sequitur ullus motus,' Thomas Bradwardine, *Tractatus de proportionibus velocitatum in motibus*, ed. H. Lamar Crosby, Jr (Madison: Univ. Wisconsin Press, 1961), 114; and 'Ab aequali aut minore vi, quam sit impedimentum, non fit motus' (Cardano, *Opus novum de proportionibus*, prop. 39, in *Opera*, 4:479–80). On the general relation between moving powers and resistances see Maier, *Die Vorläufer Galileis*, 53–72; tr. Sargent, *On the Threshold of Exact Science*, 40–60; and Clagett, *Science of Mechanics*, 421–503.

20 The point of this frivolous objection to the axiom would seem to be that action is entirely the wrong sort of thing to arise from the abstraction *equality* itself, rather than from an equal *force*. I do not know Moletti's source for this objection.

21 Cf. pseudo-Aristotle, *Quaestiones mechanicae*, qu. 7, 851b6–14; tr. Leonico, f. xxxv recto; while question 7 concerns only the relation between the trim of the sails and the rudder when sailing on the wind, in his note to it Leonico adds what becomes Moletti's main point: 'Oportet autem non vehementem esse ventum, ut temonem audiat nauis' (f. xxxv recto).

22 Cf. Tartaglia, who compares the performance of stone, lead, and iron cannonballs in *Quesiti*, 2, ff. 31v–36v.

23 Moletti himself was born in Messina, Sicily, and spent some time in Naples before coming to Venice: see Aloysius Lollinus, *Vite*, Belluno, Biblioteca Civica MS. 505. Cart. S. XVII, f. 75v.

24 Literally, 'it would not be pushed forward.'

25 Moletti's point is that walls and salients faced with hard stone tend to shatter and collapse when bombarded, while those faced with lighter materials or protected by soft materials such as straw, bales of wool, or baskets of earth tend to absorb the impact or allow the balls to pass through into the rubble or earth fill. Tartaglia comes to much the same conclusion when he suggests that ships sustain less impact from cannonballs than fixed targets because they yield to the blow by moving (*Quesiti*, 1, qu. 16, f. 22r). On the use of earthworks and other materials in fortifications, see Simon Pepper and Nicholas Adams, *Firearms and Forifications: Military Architecture and Siege Warfare in Sixteenth-Century Italy* (Chicago: Univ. Chicago Press, 1986), 72–6. On the problem of impact in general and for a discussion of Moletti's and Tartaglia's contributions, see W.R. Laird, 'Patronage of Mechanics and Theories of Impact in Sixteenth-Century Italy,' in *Patronage and Institutions: Science, Technology and Medicine at the European Court, 1550–1750*, ed. Bruce T. Moran (Woodbridge: Boydell Press, 1991), 51–66, esp. 58–61.

26 This does not come up again in the *Dialogue*.

27 Pseudo-Aristotle, *Quaestiones mechanicae*, qu. 34, 858a23–b3; tr. Leonico, f. liii verso.

28 See Aristotle, *Physics*, 8.10, 266b27–267a12; the alternative alludes to impetus theory: see Anneliese Maier, *Zwei Grundprobleme der scholastischen Naturphilosophie*, 2nd ed. (Rome: Storia e Letteratura, 1951); and 'Die naturphilosophie Bedeutung der scholastischen Impetustheorie,' *Scholastik* 30 (1955): 321–43; rpt. in Maier, *Ausgehendes Mittelalter* (Rome: Storia e Letteratura, 1964), 1:353–79; tr. Sargent, *On the Threshold of Exact Science*, 76–102; Clagett, *Science of Mechanics*, 505–40; and Ernest Moody, 'Galileo and His Precursors, ' in *Studies in Medieval Philosophy, Science, and Logic* (Berkeley: Univ. California Press, 1976), 393–408.

29 Pseudo-Aristotle, *Quaestiones mechanicae*, qu. 32 and 33, 858a14–23; tr. Leonico, f. liii recto–verso.

30 Cf. Aristotle, *Physics*, 4.8, 215b10–12; 'The heavier the medium through which a body passes, the more difficult is its descent in passing through it' (Jordanus de Nemore, *De ratione ponderis*, R4.02, ed. and tr. Moody and Clagett, *The Medieval Science of Weights*, 212–13); and 'Omne mobile naturaliter motum, seu violenter velocius movetur in medio rariore, quam densiore,' Girolamo Cardano, *Opus novum de proportionibus*, prop. 32, *Opera*, 4:477. Beginning with this paragraph, the rest of the text was copied from MS. S by Venturi into what is now MS. G and subsequently printed by Caverni (see the Introduction, 20–1 above). The first portion of it is also translated and discussed in Thomas B. Settle, 'Galileo and Early Experimentation,' 10–12.

31 This was the usual medieval interpretation of Aristotle's views, based on *Physics*, 4.8, 215a24–216a20, and *De caelo*, 1.7, 275a32–b2; see I.E. Drabkin, 'Notes on the Laws of Motion in Aristotle,' *American Journal of Philology* 59 (1938): 60–84; Ernest A. Moody, 'Galileo and Avempace: The Dynamics of the Leaning Tower Experiment,' *Journal of the History of Ideas* 12 (1951): 163–93, 375–422; rpt. in Moody, *Studies in Medieval Philosophy, Science, and Logic* (Berkeley: Univ. California Press, 1975), 203–85; Moody, 'Galileo and His Precursors,' in *Studies in Medieval Philosophy, Science, and Logic*, 393–408; James A. Weisheipl, 'Motion in a Void: Aquinas and Averroes,' in *St. Thomas Aquinas Commemorative Studies*, 2 vols, ed. A.A. Maurer (Toronto: Pontifical Institute of Mediaeval Studies, 1974), 1:467–88; Edward Grant, 'Aristotle, Philoponus, Avempace, and Galileo's Pisan Dynamics,' *Centaurus* 11 (1965): 79–95; rpt. in Grant, *Studies in Medieval Science and Natural Philosophy* (London: Variorum, 1981).

32 Moletti's assertion that bodies of different weights and materials fall together, and the implication that he performed tests himself to show it, are his great claim to fame; see Caverni, *Storia*, 4:271; Favaro, 'Amici e corrispondenti de Galileo Galilei: XL. Giuseppe Moletti,' 89; Settle, 'Galileo and Early Experimentation,' 10–12. On free fall in general, see Stillman Drake, *History of Free Fall, Aristotle to Galileo* (Toronto: Wall and Thompson, 1989), esp. 1–33.

33 Aristotle, *De caelo* 1.8, 277a23–33; Cf. 'Res gravis, quo amplius descendit, eo fit descendendo velocior' (Jordanus de Nemore, *De ratione ponderis*, R4.06, tr. Moody and Clagett, *Medieval Science of Weights*, 216–17); and 'Motus naturalis in fine, hoc est eundo versus finem, non stanti quando incipit et vadit versus finem intentitur secundo (*sic, but should be* primo) modo. Patet hoc, nam aliter grave non moveretur velocitate infinita antequam veniret ad centrum si distaret distantia infinita, cuius oppositum dicit Aristoteles' (Albert of Saxony, *Questiones [subtilissime] Alberti de Saxonia in libros de celo et mundo Aristotelis* [Venice, 1492], ed. in Clagett, *Science of Mechanics*, 568; the insertion is Clagett's); 'Omnis motus naturalis in aequali medio validior est in fine, quam in principio: violentus contra' (Cardano, *Opus novum de proportionibus*, prop. 31, *Opera*, 4:477). On the possibility of an actual infinite in the Middle Ages, see Anneliese Maier, 'Diskussionen über das actuell Unendliche in der ersten Hälfte des 14. Jahrhunderts,' *Divus Thomas*, 3rd series, 24 (1947): 147–66, 317–37; rpt. in *Ausgehendes Mittelalter: Gesammelte Aufsätze zur Geistesgeschichte des 14. Jahrhunderts* (Rome: Storia e Letteratura, 1964), 1:41–85, 460–2; and John E. Murdoch, 'Henry of Harclay and the Infinite,' in *Studi sul XIV secolo in memoria Anneliese Maier*, ed. Alfonso Maierù and Agostino Paravicini Bagliani (Rome: Storia e Letteratura, 1981), 219–61.

34 Cf. Galileo's calculation of the time it would take for a cannonball to fall from

the sphere (or orbit) of the moon in the Second Day of the *Dialogue Concerning the Two Chief World Systems*, tr. Stillman Drake, 2nd ed. (Berkeley: Univ. California Press, 1967), 218–25.

35 Cf. John Buridan, *Quaestiones super libris quattuor de caelo et mundo*, ed. E.A. Moody (Cambridge, Mass.: Harvard Univ. Press, 1942), 178; tr. in Clagett, *Science of Mechanics*, 559.

36 The fallacy here lies in the implicit measurement of speed not as distance over time, but as effect or impact. In 1604 Galileo also apparently toyed with defining speed in relation to impact rather than to time – see Stillman Drake, *Galileo at Work: His Scientific Biography* (Chicago: Univ. Chicago Press, 1978), 100–2.

37 'Spheres are to one another in the triplicate ratio of their respective diameters' (Euclid, *Elements*, 12.18, tr. Heath, 3:434).

38 Girolamo Cardano, *Opus novum de proportionibus*, prop. 110, *Opera*, 4:515–16; this proposition is reprinted and translated in Lane Cooper, *Aristotle, Galileo, and the Tower of Pisa* (Ithaca: Cornell Univ. Press, 1935), 74–7.

39 According to Roger Bacon and Walter Burley, in the clepsydra the natural tendancy of heavy bodies to fall is in fact suspended to prevent the occurence of a vacuum: see Edward Grant, 'Medieval Explanation and Interpretation of the Dictum that "Nature Abhors a Vacuum,"' *Traditio* 29 (1973): 327–55; rpt. in Grant, *Studies in Medieval Science*; and Charles B. Schmitt, 'Experimental Evidence for and against a Void: The Sixteenth-Century Arguments,' *Isis* 58 (1967): 352–66.

40 On these views, see the Introduction, 35–6 above.

41 See, e.g., John Buridan, *Questions on the Four Books on the Heavens and World of Aristotle*, tr. in Clagett, *Science of Mechanics*, 558.

42 I.e., the falling body would somehow have to know where it was in order to adjust its speeds accordingly. Gasparo Cardinal Contarini (1483–1542), in his *De elementis*, attributes to 'certain physicians' the opinion that all of nature is directed by an intelligence, so that heavy bodies know to exert more effort the closer they get to their natural places (see Duhem, *Études*, 3:183).

43 Santa Barbara is the ducal church in Mantua, and Franceschino is Francesco di Rovigo, who was organist there in the 1570s. See Pierre M. Tagmann and Michael Fink, 'Rovigo, Francesco (Franceschino),' *The New Grove Dictionary of Music and Musicians* (London: Macmillan, 1980), 16:279–80; and Iain Fenlon, *Music and Patronage in Sixteenth-Century Mantua* (Cambridge: Cambridge Univ. Press, 1980), 108 (thanks to Jeffrey G. Kurtzman for these references).

44 Aristotle, *Physics*, 4.8, 215b19–22; on the vexed question of motion in void, see Edward Grant, 'Motion in the Void and the Principle of Inertia in the Middle Ages,' *Isis* 55 (1964): 265–92; rpt. in *Studies in Medieval Science and Nat-*

ural Philosophy (London: Variorum, 1981); Grant, 'Bradwardine and Galileo: Equality of Velocities in the Void,' *Archive for the History of Exact Sciences* 2 (1965): 344–64; rpt. in Grant, *Studies in Medieval Science;* James A. Weisheipl, 'Motion in a Void: Aquinas and Averroes,' in *St. Thomas Aquinas 1274–1974: Commemorative Studies,* 2 vols. (Toronto: Pontifical Institute of Mediaeval Studies, 1974), 1:467–88; rpt. in Weisheipl, *Nature and Motion in the Middle Ages,* ed. William E. Carroll (Washington: Catholic Univ. America Press, 1985), 121–42; and in general Edward Grant, *Much Ado about Nothing: Theories of Space and Vacuum from the Middle Ages to the Scientific Revolution* (Cambridge: Cambridge Univ. Press, 1981).

45 Moletti seems to have drawn from several of Cardano's propositions for the elements of this proof: in addition to the propositions already cited, see prop. 30: 'In omni corpore mobili in medio, partes medii resistunt obviae, aliae impellunt' (*Opus novum de proportionibus, Opera,* 4:477); cf. 'The greater the depth, the slower the descent' (Jordanus de Nemore, *De ratione ponderis,* R4.04, ed. and tr. Moody and Clagett, 214–15).

APPENDIX

<Il Soggetto e le Parti Della Matematica>

[continuing from p. 80 above:]

<Con tutto ciò, molti sono stati di parere, et con ragione, che la mecanica fosse più tosto matematica che naturale, essendo che se le scienze hanno da ricevere la denominatione loro dalla ragione formale, la quale essendo in questo caso la demostratione, et le demostrationi delle mecaniche essendo quelle>[1] [D, 59r] del matematico, seguirà che la mecanica sia subalternata alla matematica. À tutto quello che s' è detto si giugne il testimonio di Proclo, autore gravissimo, il quale la mette tra le parti della matematica. Come Vostra Signoria sapoi[2] le scienze nelle loro prima divisione sono di due sorti, percioche ó sono prattiche ò contemplative. Et lasciando le prattiche da parte, diremo delle contemplative le quali sono tre, cioè[3] la metafisica, la matematica, et la fisica ò naturale. Ne possono essere altre scienze oltre queste, essendo che se altre vene sono, sono dalle dette dependenti. La sufficienza della prima divisione possiamo noi venir deducendola così. È l' huomo composto di corpo et[4] di anima, et, come veggiamo, et quanto all' una parte, et quanto all' altra è imperfetto, et però era bisogno trovar vie che potessero ridurr' à perfettione l' una et altra parte, il[5] che da' savii fà fatto con l' aiuto delle scienze, così prattiche come contemplative. Il fine adonque delle scienze prattiche è stato di ridurre à perfettione l' operationi dell' huomo, et il fine delle contemplative[6] di ridurre à perfettione la parte

1 Con ... quelle *S, 297v*
2 sapoi *ins. D* poi *super* sa
3 *ante* cioè *scrip. et canc.* percioche ò e *D*
4 *ante* et *scrip. et canc.* di *D*
5 *ante* il *scrip. et canc.* peró *D*
6 *ante* contemplative *scrip. et canc. verbum illegibile D*

APPENDIX

<The Holograph Fragment on the Subject of Mathematics>

[continuing from p. 81 above:]

<With all such it has seemed to many – and with reason – that mechanics is more mathematical than natural, since if sciences receive their denomination from their formal principle, which in this case is demonstration, and the demonstrations of mechanics are those of mathematics, it follows that mechanics is subalternated to mathematics. To all this that has been said one can add the testimony of Proclus, a most weighty authority, who set mechanics among the> parts of mathematics.[1] As you know, the sciences in their first division are of two kinds, since they are either practical or contemplative.[2] And leaving the practical aside, we say that the contemplative are three in number, that is, metaphysics, mathematics, and physics or natural philosophy. Nor can there be sciences other than these, for if there are, they are dependents of these. The sufficiency of the first division we can come to deduce in this way. Man is composed of body and soul, and, as we have seen, both parts are imperfect, and so it was necessary to find ways that each could be brought to perfection, which was done by the wise with the help of the sciences, both practical and contemplative. The goal, then, of the practical sciences was to bring to perfection the works of man, and the goal of the contemplative to bring to perfection his intellective part. The distinction, therefore, between these two sciences is that the practical has work

intellettiva di quello. Et però la distinzione di queste due scienze è che la prattica ha per fine l' operazione, et la contemplativa la cognitione. Ma non ogni operatione è fine del prattico, poiche quando così fosse, l' operationi viliose et laide sarebbono fine del prattico; ma solo l' operationi virtuose et buone hanno da essere fine della scienza prattica. Et il fine essendo quello che 'l bene, et il vero et non l' apparente, però il bene agibile è fine del prattico. Così ancora non ogni cognitione[7] ha da essere il fine del contemplativo, ma solo la cognitione[8] della verità; et però la scienza contemplativa haverà[9] per fine [D, 59v] il vero. Ma perché l' operatione è di due sorti, una n' è[10] laquale doppo se lassa alcuna cosa fatta et l' altra che non lassa cosa alcuna (si come è del musico et del muratore ò marangone), però quella che non lassa opera alcuna doppo il fatto fà detta attione, et l' altro fattione. In[11] scienza che comprende l'attioni et che regola quelle, ò regola l' attioni d' un solo, ò di tutta una casa, ò di tutta una città. Se quelle d' un solo, è l' etica; se d' una casa, è la economica; se d' una città, è la politica. La scienza che versa poi intorno alle fattioni[12] è quella che regola l' arti fattive, tutte le quali sono di più sorti, come Vostra Signoria sa, et le quali tutte hanno iloro principii et iloro universali. Percioche vengon tutte fatti dall' esperienza, dalle quale vien fatta la memoria, et da questa vien fatto l' universale, <c'è> materia commune così alle arti come alle scienze. Et però[13] non picciolo obligo l' ha d' havere à coloro che prima rozzamente comincianno le arti et le scienze, essendo che senza loro non haveremo tanta copia di arti et tante scienze. Che poi tra artefici habbia da essere più stimato quello che miglior conto supra rendere dell' arte sua che uno che redia meno è chiaro da se. Come s' è detto poi il fine del contemplativo fù di conoscere il vero, ma non d' una sola specie, ma di tutte le cose che nel mondo sono. Et[14] tutte le cose possono essere considerate nella loro prima cognitione, sotto la consideratione dell' ente ò d' una cosa ch' è; però prima è una scienza che considera le cose sotto tal concetto, la quale è la metafisica, che considera l' ente. Et perché questo ente si divide ne' dieci predicamenti, però tale scienza sotto concetto universale

7 cognitione *ins.* contemplatione, quod *scrip. et canc. D*
8 *ante* cognitione *scrip. et canc.* contemplatione *D*
9 *ante* haverà *scrib. et canc.* ha da essere *D*
10 una n' è *ins. D*
11 *ante* In *scrip. et canc.* La prima comprende *D*
12 *ante* fattioni *scrib. et canc. verbum illegibile D*
13 *ante* Et però *scrib. et canc.* Percio *D*
14 *ante* Et *scrib. et canc.* però *D*

for a goal, and the contemplative has cognition. But not every kind of work is a goal of practice, since if this were so, vile and filthy works would be goals of practice; but only virtuous and good works are to be the goal of practical science. And since the goal is what is good, and a true good not a seeming good, to act well is the goal of the practical. Similarly, not every cognition has to be the goal of contemplation, but only the cognition of the truth; and therefore contemplative science will have as its goal the true. But because work is of two kinds, one that leaves behind a thing made and the other that leaves nothing (as does the musician and the bricklayer or mason), that which does not leave any work after the deed is called action, and the other is called production.[3] In science that involves actions and that rules them, it rules either the actions of an individual, a whole household, or an entire city. If those of an individual, it is ethics; if of a household, it is economics; if of a city, politics. The science that concerns production, then, rules the productive arts, all of which are of many kinds, as you know, and all of which have their own principles and universals. Since all deeds come from experience, from which memory is made, and from which in turn the universal is made, <there is> a common subject matter for the arts just as for the sciences. And thus there is no small debt owed to those who first crudely originated the arts and the sciences, since without them we would not have such copious arts and so many sciences. That among the artisans the one that excels to give a better account of his art should be more highly esteemed than one who gives less is self-evident. As was said, then, the goal of the contemplative was to know the true, but not only of one kind, but of all the things that are in the world. And all things can be considered in their first cognition, under the consideration of being, or of one thing that is; therefore first is a science that considers things under such a concept, i.e., metaphysics, which considers being. And because being is divided into the ten categories, metaphysics considers being under the universal concept and all its parts, such as substance, quantity, quality, relation, where (or place), when (or time), to be

considera l' ente et tutte le parti sue, come la sostanza, la quantità, la qualità, la relazione,[15] dove (ò il luogo), quando (ò il tempo), l' essere[16] collocato (ò stare in alcun modo), havere (come essere vestito, calzato,[17] ò armato), fare (come cucire, tagliare), patire (come essere indicato). [D, 60r] Et perché la sostanza ó è generabile et corruttibile ó eterna, peró il considerare minutamente le proprietà tutte della sostanza generabile et corruttibile, et così dell' eterna, come congionta con la detta, non partene[18] al metafisica, che fosse secondo il suo primo proponimento, però ne lasciò il carico al fisico ò al naturale. Ma perché con la sostanza corporea, ó eterna ó corruttible ch' ella sia, è[19] congionta la quantità continua et il numero, il quale è ancora congionto con le sostanze astratto. Et il continuo et il discreto hanno molte proprietà disgiunte dalla contemplatione così naturale come metafisica, perciò[20] s' è lassato il carico di considerare le sudette proprietà al matematico. Ne d' altro predicamento s' è potuto fare particolare scienza, poiche ciascun altro ha il suo determinato soggette dove s' ha da ritrovare. Il che non avviene al quanto essendo ch' è atto à ritrovarsi in tutte le sorti delle sostanze et in tutti i soggetti, essendo che 'l continuo (come fà detto) è così nelle sostanze eterne corporee[21] come corruttibili, et il numero è così nelle astratte come nell' altre tutte. Ma perché era bisogno à filosofi dare ad intendere queste divisioni per altri termini, per farne gli addiscenti da principio più capaci, però vennero dimostrando questa divisione per altre vie et presero per mezo la definitione.[22] Il metafisico diffinisce le cose sue senza materia et senza movimento, essendo tale l' essere delle cose ch' egli considera. Il fisico diffinisce[23] le cose per la materia et per il movimento, essendo che le cose considerate da lui sono così conditionate. Il matematico diffinisce le cose remote dalla materia, non secondo l' essere ma per opera dell' intelletto, poiche veramente non si può ritrovare sorte alcuna di quantità che non sia congionta con alcuna sorte di sostanza. [D, 60v] Ma la quantità continua non può ritrovarsi mai disgiunta dalla materia sensibile, et però hanno detto i filosofi questa hav-

15 la quantità, la qualità, la relazione, *ins. D* la qualità, la quantità, la relazione, *quod canc.*
16 *ante* l' essere *scrib. et canc.* et *D*
17 calzato] cazato *D, ante quod scrib. et canc.* ó
18 partene *lectio incerta D*
19 *ante* è *scrib. et canc.* et così con le sostanze astratte, et incorpor é *D*
20 *ante* perciò *scrib. et canc.* et perché *D*
21 corporee *ins. D*
22 definitione] definione *D*
23 diffinisce] diffisce *S*

located (or to stand in some way), *habitus* (e.g., to be dressed, shod, or armed), to do (e.g., to sew, to cut), to suffer (e.g., to be pointed out).[4] And although substance is either generable and corruptible or eternal, nevertheless to consider in detail the properties of all generable and corruptible substances, as well as those of eternal substances, as conjoined to them, does not pertain to the metaphysician according to his first purpose, but the job was left to the physicist or natural philosopher. But since continuous quantity and number are conjoined with corporeal substance, whether it be eternal or corruptible, they are also conjoined with abstract substances. And since the continuous and the discrete have many properties separate from natural as well as from metaphysical contemplation, the job of considering the aforesaid properties was left to the mathematician. Of no other category was it necessary to make a separate science, since each of the others has its definite subject where it must be found. This does not happen in the case of quantity, since it is found in all kinds of substance and in all subjects, since the continuous (as was said) is as much in eternal corporeal substance as in corruptible, and number is as much in the abstract as in all others. But because it was necessary for philosophers to give to the understanding these divisions in other terms, in order to be better able to make the descent from principles, they came to demonstrate this division through other ways, and took definition as a means. The metaphysician defines his things without matter and without movement, because such is the being of the things he considers. The physicist defines things through matter and through movement, because the things considered by him are qualified in this way. The mathematician defines things separate from matter, not in being but by an act of the intellect, for one cannot in reality find any sort of quantity that is not conjoined with some sort of substance. But continuous quantity can never be found separated from sensible matter, and thus the philosophers said that it needs to have position and it also involves an intelligible matter, which according to them is an indeterminate continuous quantity. But because the discrete can be found outside

ere di bisogno di positione et però ancora intesero una materia intelligibile, che viene ad essere secondo loro la quantità continua indeterminata. Ma perché la discreta si può trovare fuori delle sostanze sensibili, poiche 'l numero si ritrova nelle sostanze separate, però questa non ha di bisogno nella contemplatione[24] sua di positione. Et questo è quello c' ha detto Aristotele nel primo della *Posteriora <analytica>* nel testo 42, che l' aritmetica era più nobile della geometria, poich' era di soggetto più astratto che la geometria, essendo che l' unità è senza positione et il ponto è con positione.

A. Mi ricordo haver sentito disputar molto ad alcuni dottori miei vassalli intorno alla divisione di queste scienze, et chi teneva la matematica scienza, et chi no. Ma insomma eran sempre, et con la ragione et l' intentà d' Aristotele, convinti coloro che tenevano la matematica non essere scienza.

P. Se questo fosse il suo proprio luogo et se Vostra Signoria non ne fosse già informato, io tenterei di addurne molte ragioni, con le quali dimostrerei la matematica essere veramente scienza. Ma poiche Vostra Signoria n' è quasi come risoluto, et ancora per non esser questo il vero luogo suo, passerò oltre. Come diceva, ha per soggetto la matematica la quantità in universale, da dove et Aristotele et Proclo hanno detto essere una scienza matematica che considera le proprietà della quantità in universale, et lequali così convengono alla continua quantità come alla discreta. Si divide poi la quantità in continua et in discr[D, 61r]eta, da dove nascono le due scienze matematiche, cioè l' aritmetica et la geometria, le quali sono le vere et assolute scienze matematiche.

A. Et s' è così come l' Altezza Vostra ha detto, ond' è c' hanno[25] messo sotto la matematica l' astronomia, la musica, la perspettiva, la mecanica, et altre? poiche considerano le cose nella materia sensibile et col movimento ò non da quello disgiunte?

P.[26] Al dubbio di Vostra Signoria sarei venuto già, et posso à quello rispondere in due modi. Il primo de' quali è secondo Gemino, citato da Proclo, il quale voleva la matematica ò considerarsi al tutto et per tutto remota dal senso, ò da quello non totalmente disgiunta. Se al primo modo, le parti della matematica saranno solamente due, cioè l' aritmetica et la geometria. Se non disgiunta dal senso et dalle cose sensibili, saranno le parti sue sei, cioè l' astronomia, la mecanica, la perspettiva, la

24 *ante* contemplatione *scrib. et canc.* positione *D*
25 c' hanno] c' hano *D*
26 *ante* P. *scrip. et canc.* A. *D*

sensible substance, since number is found in separated substances, nevertheless it does not have the need of position in its contemplation. And this is what Aristotle said in *Posterior Analytics*, book 1, textus 42, that arithmetic was more noble than geometry, since it was of a subject more abstract than geometry, because the unit is without position and the point has position.[5]

A. I remember having heard much dispute among some doctors who were my vassals, over the division of this science, some holding mathematics to be a science and some not. But in the end they were always convinced, both by reason and by the opinion of Aristotle, by those who held that mathematics was not a science.[6]

P. If this were the proper place and if you were not already informed, I would try to give many reasons by which I would show that mathematics truly is a science. But since you are already almost convinced, and again because this is not the proper place for it, I shall pass on to other things. As you said, mathematics has for its subject quanitity in the universal, whence both Aristotle and Proclus said that mathematics was a science that considers the properties of quantity in the universal, both those that belong to continuous quantity and to discrete. Quantity is divided into continous and discrete, whence arise the two mathematical sciences, i.e., arithmetic and geometry, which are the true and absolute mathematical sciences.[7]

A. And if it is as you have said, how is it that they have put astronomy, music, optics, mechanics, and others under mathematics? – for they consider things in sensible material and with movement or not separate from it.

P. I was about to come to this doubt of yours, and I can answer it in two ways. The first is that of Geminus, as cited by Proclus, who wanted mathematics to be considered either completely and utterly remote from sense, or not totally separated from it. If in the first way, the parts of mathematics would be only two, i.e., arithmetic and geometry. If not separated from sense and from sensible things, its parts would be six, i.e., astronomy, mechanics, optics, geodesy, canonics, and calculation.[8] The second way I can respond is that of Aristotle, who from natural phi-

geodesia, la canonica, et la supputatrice. Il secondo modo con quale posso rispondere è quello d' Aristotele, il quale vien dalla naturale et dalla matematica, facendo alcune scienze che le chiama mezane, sicome sono le sopradette, percioche dal naturale ne pigliano la materia et dal matematico ne pigliano il modo di dimostrare. Et questo modo d' Aristotele convene quasi allo stesso segno con quello di Gemino detto di sopra. Diró adonque l' astronomia essere scienza media secondo Aristotele, tra la naturale et la matematica, percioche dal naturale ne piglia la materia ch' è il cielo et il movimento di quello, et del matematico ne piglia i principii et il modo di demonstrare et di supputare. Quel c' ho detto dell' astronomia, diró della musica, della perspettiva, et della mecanica.

A. Mi sovviene appunto haver sentito far molte dispute se l' astronomia poteva essere scienza da [D, 61v] per se, et così l' altre dette di sopra, perché diceva il contradicente, il cielo tutto, insieme col movimento suo, vien considerato da filosofo naturale, adonque non lassa parte che possa considerarsi da altro scientifico. Diceva ancora se l' astronomia fosse scienza distinta dalla fisica, adonque seguirebbe che non fossero tre le scienze contemplative, perché ha da essere l' astronomia, ó nelle contemplative ó nelle prattiche: non <è> nelle prattiche, poich' è stata numerata tra le contemplative, cioè la matematica, adonque non sono tre le contemplative. Oltre queste ragioni ne allegava molte altre, ma queste erano stimate le principali.

P. Stimo che molto s' ingannasse il contradicente nominato da Vostra Signoria. Percioche quelle scienze son' una c' hanno il soggetto uno et che provano le stesse proprietà dello stesso soggetto con gli stessi principii; ma così non avviene al naturale et all' astronomo. Poiche si come il metafisico ha lasciato la cura di considerare le proprietà della sostanza corporea così generabile come ingenerabile al fisico, da dove n' è vala la seconda scienza; così il naturale ha lassato il considerare le proprietà del movimento de' cieli all' astronomo. Considera il naturale il movimento del cielo come consecutivo al corpo celeste, et lo considera circulare, ma non sotto ragione di numerabile, si come fa l' astronomo. Et questa ragione di numerabile fa distinta la scienza dell' astronomia dalla naturale. Appresso considera il naturale molte proprietà del movimento del cielo, ma non con quella minuteza che fa l' astronomo: ne si cura non

losophy and mathematics makes certain sciences that he calls 'middle,' which are those listed above, since they take their material from the natural philosopher and their way of demonstrating from the mathematician. And Aristotle's way comes to almost the same thing as that of Geminus mentioned before. So I shall say that astronomy is a middle science according to Aristotle, in between the natural and the mathematical, since it takes its material, i.e., the heavens and their movements, from the natural, and from the mathematician it takes the principles and the mode of demonstrating and calculating. What I have said about astronomy also applies to music, optics, and mechanics.[9]

A. I remember distinctly having heard many disputes over whether astronomy could be a science in itself, and also over the other sciences mentioned above. For the one denying <that astronomy is a science> argued that since the whole of heaven, together with its movement, is treated by the natural philosopher, there is no part left over that can be treated by another scientist. Again, he said that if astronomy were a science distinct from physics, then it would follow that there would not be <only> three contemplative sciences. For astronomy has to be either among the contemplative or the practical: <it is> not among the practical, since it was counted among the contemplative, i.e., the mathematical. Therefore there are not <just> three contemplative sciences. Beyond these reasons he adduces many others, but these are thought to be the main ones.[10]

P. I think that the one you refer to who denies <that astronomy is a science> is greatly mistaken. For those sciences are one that have one subject and that prove the same properties of the same subject with the same principles; but this is not the case for the natural philosopher and for the astronomer. For just as the metaphysician has left the job of considering the properties of corporeal substance, both generable and ingenerable, to the physicist, whence the second science <i.e., physics> arose; so the natural philosopher has left it to the astronomer to consider the properties of the movement of the heavens. The natural philosopher considers the movement of the heavens as consequent to the celestial body, and he considers it to rotate, but not under the aspect of the numerable, as does the astronomer. And this aspect of the numerable makes the science of astronomy distinct from natural philosophy. Further, the natural philosopher considers many properties of the move-

solo d' osservarle ma ne ancora di predirle in futuro, fini particolari dell' astronomo. Queste differenze adonque sono quelle [D, 62r] che fanno distinta l' astronomia della fisica. Oltre à ciò quando Vostra Signoria dice che non sarebbono tre le scienze ma più, rispondo che le principali sone tre, ma non siegue che non ne possano essere altre à quelle subalternate et da quelle dependenti. Ben dirà Vostra Signoria non poter ritrovarsi soggetto che non sia compreso dalle dette tre scienze; ma che uno pigli una parte d' uno di quei soggetti et che ne venghi considerando le particolari proprietà, non so perché possa vietarsi, si come fa l' astronomo, et il perspettivo, et gli altri detti di sopra. Oltra di questo, non so come quel contradicente haverebbe rispecto à questa ragione. Ciascuna volta ch' una facoltà ha quelle cose che si richieggono ad una scienza è per consequente tale; ma l' astronomia ha quelle cose che si richieggono ad una scienza; adonque è tale. Nella scienza si richiede il soggetto, del quale se ne hanno da considerare le proprie passioni, et s' hanno da dimostrar del soggetto per i suoi principii. Che l' astronomo habbia il soggetto è chiaro, poiche 'l suo soggetto è il cielo et il movimento di quello. Dimostra di tal soggetto le proprie passioni col mezo de' suoi principii.

A. Et in questo diceva il contradicente consistere l' inganno dell' astronomo, cioè che crede haver soggetto, del quale ne dimostri le passioni, et così non è, essendo tutta via il cielo et il movimento suo soggetto del naturale.

P. Darò à Vostra Signoria un' essempio, con il quale la farò capace del tutto. Dicami, quel che fa[27] i corzaletti et le celate, intorno à che si travaglia?

A. Intorno al ferro.

P. Bene. Et quel che fa i vomeri et le zappe, c' ha per soggetto?

A. Parimente il ferro.

P. Queste due arti sono [D, 62v] le stesse, ò sono tra diloro differenti?

AN. Diró che sono differenti, essendo uno il fino di chi fa i corzaletti et un' altro di chi fa le zappe. Colui che fa i corzaletti ha per fine d' introdur nel forma del corzaletto, et quel che fa i vomeri d' introdurre nella stessa materia la forma del vomero, le quali forme sono differenti.[28] Et à far ciò l' uno et altro diversamente preparano la materia, percioche[29] il

27 fa *ins. D*
28 le quali forme sono differenti *ins.* D
29 *ante* percioche *scrib. et canc.* loro *D*

ment of the heavens, but not in the detail of the astronomer: the astronomer not only observes the movements of the heavens but also predicts them into the future, the particular goal of the astronomer. These differences then are such that make astronomy distinct from physics.[11] Beyond this, when you say that there would be not three sciences but more, I reply that there are three principal sciences, but that it does not follow that there cannot be others subalternated to and dependent on them. You spoke well that there cannot be found a subject that is not included within these three sciences; but that one takes one part from one of those subjects and comes to consider its particular properties, I can see no reason to forbid, just as astronomy, and optics, and the others mentioned above do. Beyond this, I do not know what the one who denies <that astronomy is a science> would reply to the following argument. Whenever a faculty has what is required for a science it is consequently one; but astronomy has what is required for a science; therefore it is one. A science requires a subject, of which it must consider the proper accidents, and it must demonstrate them of the subject through its own principles. That the astronomer has a subject is clear, since his subject is the heavens and their movement. He demonstrates the proper accidents of this subject by means of his own principles.

A. And in this, the one denying said, consists the error of the astronomer, i.e., that he believes he has a subject of which he demonstrates the accidents, and this is not so, since the whole path of the heavens and their movement is the subject of the natural philosopher.

P. I shall give you an example with which I shall make everything possible. Tell me, those who make armour, what do they work with?

A. With iron.

P. Good. And those who make ploughs and hoes, what do they have for a subject?

A. Similarly iron.

P. Are these two arts the same, or are they different from each other?

AN. I shall say that they are different, since the one that makes armour has one goal, and the one that makes hoes another. The one that makes armour has for a goal to impose the form of armour, and the one that makes ploughs to impose on the same material the form of the plough, and these forms are different. And to make them they each work their material differently, in that the first flattens it into sheets and

primo la distende in lame et il secondo così non fa, oltre à molte altre cose in che differiscono nel lavorare. Ma non mi pare che Vostra Altezza possa dimostrare l' istesso dell' astronomia et della naturale.

PR. Anzi[30] l' istesso dimosterò. Vostra Signoria ha da sapere che le scienze contemplative convengono tutte in un fine universale, il quale è di conoscere la verità. Ma la differenza loro consiste ne' fini particolari. Ha per fine l' astronomo la cognitione della verità, et così il naturale.

30 *ante* PR. Anzi *scrib. et canc.* PR. Anzi *D*

the second does not, among many other things in which they differ in their work. But I do not think that you can demonstrate the same thing of astronomy and natural philosophy.

PR. In fact I will demonstrate the same thing. You must know that the contemplative sciences all share a universal goal, which is to know the truth. But their difference consists in their particular goals. The astronomer has for a goal the cognition of the truth, just as the natural philosopher has.

NOTES TO THE APPENDIX

1 Proclus, *In primum Euclidis Elementorum librum commentarii*, ed. Friedlein, 38–41, tr. Morrow, 31–4.

2 Moletti here is following the medieval standard division of the sciences: see James A. Weisheipl, 'Classification of the Sciences in Medieval Thought,' *Mediaeval Studies* 27 (1965): 54–90; and Roger Ariew, 'Christopher Clavius and the Classification of Sciences,' *Synthese* 83 (1990): 293–300. See also Moletti's discussion of the general divisions of science in the *Discorso*, ff. 122v–125r.

3 Cf. Aristotle, *Nicomachean Ethics*, 1.1, 2, 1094a1–b11.

4 These are the ten categories of Aristotle, which Moletti understands in the common way as categories of being; see Aristotle, *Categories*, 4, 1b25–2a4.

5 Aristotle, *Posterior Analytics* 1.27, 87a31–7: 'Scientia autem est certior et prior scientia, quae ipsius quod et propter quid est eadem, sed non ea, quae seorsum ipsius quod ab ea quae propter quid: et quae non de subiecto est ea quae de subiecto, ut arithmetica harmonica; et quae ex paucioribus ea quae est ex appositione, ut arithmetica geometria. Dico autem ex appositione, ut unitas est substantia sine positione, punctus vero substantia cum positione, hoc autem ex appositione,' *Aristotelis Stagiritae posteriorum resolutoriorum libri duo cum Averrois Cordubensis magnis commentariis*, I. tex. 42, in *Aristotelis omnia quae extant opera. Averrois in ea opera omnes commentarii*, 11 vols. in 13 (Venice: Juntas, 1560–2; rpt. 1573–4), I. pars II, f. 374 A–B.

6 The status and certitude of mathematics were hotly debated in the sixteenth century: see the series of studies by Giulio Cesare Giacobbe, 'Il *Commentarium de certitudine mathematicarum disciplinarum* di Alessandro Piccolomini,' *Physis* 14 (1972): 162–93; 'Francesco Barozzi e la *Questio di certitudine mathematicarum*,' *Physis* 14 (1972): 357–74; 'La riflessione metamatematica di Pietro Catena,' *Physis* 15 (1973): 178–96; 'Alcune cinquecentine riguardanti il processo di rivalutazione epistemologica della matematica nell'ambito della rivoluzione scientifica rinascimentale,' *La Berio* 13 (1973): 7–44; 'Epigoni nel Seicento della "*Quaestio de certitudine mathematicarum*": Giuseppe Biancani,' *Physis*, 18 (1976): 5–40; 'Una gesuita progressista nella "Questio de certitudine mathematicarum" rinascimentale: Benito Pereyra,' *Physis* 19 (1977): 51–86. Moletti himself contributed to the debate: see his incomplete *Rudimenta quedam pro mathematicis disciplinis*, Milan, Bibli. Ambr. MS. D 442 inf., ff. 1r–19v, which describes the division of the sciences, mathematical abstraction, and mathematical demonstration; on this work see Adriano Carugo, 'Giuseppe Moleto: Mathematics and the Aristotelian Theory of Science at

Padua in the Second Half of the 16th-Century,' in *Aristotelismo Veneto e Scienza Moderna*, ed Luigi Olivieri, 2 vols., Saggi e Testi 17–18 (Padua: Antenore, 1983), 1: 509–17. See also Moletti's discussion of the subject of mathematics and mathematical abstraction in the *Discorso*, ff. 125r–126r.

7 Cf. Moletti, *Discorso*, ff. 126r–v.

8 Proclus, *In primum Euclidis Elementorum librum commentarii*, ed. Friedlein, 38, tr. Morrow, 31; this is a close paraphrase of Proclus.

9 For Aristotle's distinction between astronomy and natural philosophy, see *Physics*, 2.2, 193b22–194a12; on the middle sciences, see the Introduction, 8–9 above, and Moletti, *Discorso*, ff. 141v–142r.

10 I have not found Moletti's source for these puerile arguments.

11 Cf. Moletti's discussion of astronomy in the *Discorso*, ff. 142r–150v.

Glossary

I have listed below those words that might be obscure to a modern reader, followed by their equivalent in modern Italian or an English translation; I have not listed words that differ only by having a single rather than a doubled letter, e.g., *mezo* (= *mezzo*), or by an accent. In a few cases I have given the Latin cognate.

adonque, adunque	dunque
augmentare	aumentare
balla	palla
constituire	costituire
cortello	coltello
dimanda	domanda
dimostrare	mostrare
et	e, ed
giuoco	gioco
haverà	avrà
hora	ora
inanti	innanzi
intalio	intaglio
iscambio	invece
lieva	leva
maraviglia	meraviglia
medietà	metà
movere	mouvere
nomano	contrapeso
onza	oncia
partene	appartene
percioche	perciò che

però	dunque
picciolo	piccolo
pingere	spingere
pinta	spinta
ponto	punto
pruova	prova
puoco	poco
se bene	*for if indeed; not* sebbene
si come	siccome
solo	suolo
sparta	fulcro; *cf. Latin* spartum
statera	stadera
temone	timone
tollere	togliere
vecte	*Latin* vectis, *lever*
venghi	venghi, venga

Bibliography

Manuscripts

Lollinus, Aloysius. *Vite: Josephus Moletius messanensis.* Belluno, Biblioteca Civica MS. 505. cart. s. XVII, ff. 75v–77v.

Mantua, Archivio di Stato, Archivio Gonzaga, serie E. XLV. 3, buste 1511, 1512, 1513, and 1514. (Letters of Giuseppe Moletti to Mantua)

Mantua, Archivio di Stato, Archivio Gonzaga, serie F. II. 8, buste 2612 and 2617. (Letters of Giuseppe Moletti to Mantua)

Milan, Biblioteca Ambrosiana MS. A 71 inf. (Letters of Giuseppe Moletti)

Milan, Biblioteca Ambrosiana MS. D 235 inf. (Fragmentary works of Giuseppe Moletti)

Milan, Biblioteca Ambrosiana MS. D 332 inf. (Letters of Giuseppe Moletti)

Milan, Biblioteca Ambrosiana MS. D 442 inf. (Giuseppe Moletti, works, including *Rudimenta quedam pro mathematicis disciplinis,* ff. 1r–19v)

Milan, Biblioteca Ambrosiana MS. S 80 sup. (Letters of Giuseppe Moletti)

Milan, Biblioteca Ambrosiana MS. S 100 sup. (Giuseppe Moletti, lectures and works, including *In librum Mechanicorum Aristolelis expositio,* ff. 154r–210v; *Alcune memorie in materia d'artiglieria* [= *Dialogue on Mechanics*] ff. 294r–318r)

Milan, Biblioteca Ambrosiana MS. S 103 sup. (Giuseppe Moletti, works, including *Discorso, che cosa sia matematica,* ff. 122r–175v)

Milan, Biblioteca Ambrosiana MS. S 105 sup. (Letters of Giuseppe Moletti)

Vatican City, Biblioteca Apostolica Vaticana MS. Vat. lat. 6194, f. 416. (Giuseppe Moletti's letter to Cardinal Sirleto)

Vatican City, Biblioteca Apostolica Vaticana MS. Vat. lat. 6195, ff. 10, 12, 14, 102, 562, 680, and 684 (Giuseppe Moletti's letters to Cardinal Sirleto)

Printed Works

Alberti, Leon Battista. *L'Architettura [De re aedificatoria]*. Ed. and tr. Giovanni Orlandi. 2 vols. Milan: Edizioni il Polifilo, 1966.
Albertus Magnus. *Posteriora analytica. Opera omnia*, 2. Ed. August Borgnet. 38 vols. Paris: Louis Vivès, 1890–9.
Allard, Guy H. 'Les arts mécaniques aux yeux de l'idéologie médiévale.' In *Les arts mécaniques au moyen âge*, ed. G.H. Allard and S. Lusignan. Montreal: Bellarmine; Paris: Vrin, 1982. Pp. 9–31.
Apostle, Hippocrates George. *Aristotle's Philosophy of Mathematics*. Chicago: Univ. Chicago Press, 1952.
Ariew, Roger. 'Christopher Clavius and the Classification of Sciences.' *Synthese* 83 (1990): 293–300.
Aristotle. *De incessu animalium*. Tr. Niccolò Leonico Tomeo, in *Opuscula nuper in lucem aedita*. Venice, 1525.
– *De motu animalium*. Tr. Niccolò Leonico Tomeo, in *Opuscula nuper in lucem aedita*. Venice, 1525.
– *Aristotelis omnia quae extant opera. Averrois in ea opera omnes commentarii*. 11 vols. in 13. Venice: Juntas, 1560–2. Rpt. 1573–4.
– *De motu animalium*. Ed. and tr. E.S. Forster. London: Heinemann; Cambridge, Mass.: Harvard Univ. Press, 1959.
Baldi, Bernardino. *Mechanica Aristotelis problemata exercitationes*. Mainz, 1621.
Bertoloni Meli, Domenico. 'Guidobaldo dal Monte and the Archimedean Revival.' *Nuncius* 7.1 (1992): 3–34.
Bertolotti, A. *Architetti, ingegneri, e matematici in relazione coi Gonzaga, Signori di Mantova, nei secoli xv, xvi, e xvii*. Genova, 1889.
Boyer, Marjorie Nice. 'Pappus Alexandrinus.' In *Catalogus translationum et commentariorum: Medieval and Renaissance Latin Translations and Commentaries*, ed. Paul Oskar Kristeller and F. Edward Cranz. 7 vols. to date. Washington: Catholic Univ. of America Press, 1960– . 2:205–13, 3:426–431.
Brown, Joseph E. 'The *Scientia de Ponderibus* in the Later Middle Ages.' PhD diss., Univ. Wisconsin, 1967.
– 'The Science of Weights.' In *Science in the Middle Ages*, ed. David C. Lindberg. Chicago: Univ. Chicago Press, 1978. Pp. 179–205.
Camerota, Michele. *Gli Scritti 'De motu antiquiora' di Galileo Galilei: Il Ms Gal 71*. Cagliari: CUEC Editrice, 1992.
Cardano, Girolamo. *Opus novum de proportionibus*. Basel, 1570. Rpt. in *Opera*. London, 1663; rpt. in facsimile New York: Johnson Reprint, 1967. Vol. 4.
Carmody, Francis J. 'Autolycus.' In *Catalogus translationum et commentariorum: Medieval and Renaissance Latin Translations and Commentaries*, ed. Paul Oskar

Kristeller, F. Edward Cranz, and Virginia Brown. 7 vols. to date. Washington: Catholic Univ. of America Press, 1960– . 1:167–8.

Carugo, Adriano. 'Giuseppe Moleto: Mathematics and the Aristotelian Theory of Science at Padua in the Second Half of the 16th Century.' In *Aristotelismo Veneto e Scienza Moderna*, ed. Luigi Olivieri. 2 vols. Saggi e Testi 17–18. Padua: Antenore, 1983. 1:509–17.

– 'L'Insegnamento della matematica all' Università di Padova prima e dopo Galileo.' *Storia della Cultura Veneta*, ed. Girolamo Arnaldi and Manlio Pastore Stocchi. 4 vols. in 7. Venice: Neri Pozza Editore, 1976–86, 4/II (1984): 151–99.

Catena, Pietro. *Universa loca in logicam Aristotelis in mathematicas disciplinas*. Venice, 1556.

Caverni, Raffaello. *Storia del Metodo Sperimentale in Italia*. 6 vols. Florence, 1891–1900; rpt. New York: Johnson Reprint, 1972.

Clagett, Marshall. *Giovanni di Marliano and Late Medieval Physics*. New York: Columbia Univ. Press, 1941.

– 'Three Notes: The *Mechanical Problems* of Pseudo-Aristotle in the Middle Ages, Further Light on Dating the *De curvis superficiebus Archimenidis*, Oresme and Archimedes.' *Isis* 48 (1957): 182–3.

– 'The Works of Francesco Maurolico.' *Physis* 16 (1974): 149–98.

– *Archimedes in the Middle Ages*. 5 vols. Vol. 1, Madison: Univ. Wisconsin Press, 1964. Vols. 2–5, Philadelphia: Memoirs of the American Philosophical Society, 117, 125, 137, 157. 1976, 1978, 1980, 1984.

Clagett, Marshall, ed. *The Science of Mechanics in the Middle Ages*. Madison: Univ. Wisconsin Press, 1959.

Cooper, Lane. *Aristotle, Galileo, and the Tower of Pisa*. Ithaca: Cornell Univ. Press, 1935.

Copernicus, Nicholas. *De revolutionibus orbium coelestium*. Ed. Jerzy Dobrzychi, tr. Edward Rosen, in *Nicholas Copernicus Complete Works*, 3 vols. Warsaw / Cracow: Polish Academy of Sciences, 1972–85. Vol. 2.

De Bellis, D. 'Nicolò Leonico Tomeo, interprete di Aristotele naturalista.' *Physis* 17 (1975): 71–93.

de la Croix, H. 'The Literature on Fortification in Renaissance Italy.' *Technology and Culture* 4 (1963): 30–50.

del Monte, Giudobaldo. *Liber mechanicorum*. Pesaro, 1577.

– *Le mechaniche*. Tr. Filippo Pigafetta. Venice, 1581.

Drabkin, I.E. 'Notes on the Laws of Motion in Aristotle.' *American Journal of Philology* 59 (1938): 60–84.

– 'A Note on Galileo's *De motu*.' *Isis* 51 (1960): 271–7.

Drabkin, I.E., and Stillman Drake. *Galileo Galilei on Motion and Mechanics*. Madison: Univ. Wisconsin Press, 1960.

Drachmann, A.G. *The Mechanical Technology of Greek and Roman Antiquity*. Copenhagen: Munksgaard, 1963.
Drake, Stillman. 'Galileo Gleanings, V: The Earliest Version of Galileo's *Mechanics*.' *Osiris* 13 (1958): 262–90.
– 'Medieval Ratio Theory vs Compound Medicines in the Origins of Bradwardine's Rule.' *Isis* 64 (1973): 67–77.
– *Galileo at Work: His Scientific Biography*. Chicago and London: Univ. Chicago Press, 1978.
– *History of Free Fall, Aristotle to Galileo*. Toronto: Wall and Thompson, 1989.
Drake, Stillman, and I.E. Drabkin. *Mechanics in Sixteenth-Century Italy*. Madison: Univ. Wisconsin Press, 1969.
Duhem, Pierre. *Les origines de la statique*. 2 vols. Paris: A. Hermann, 1905–6.
– *Études sur Léonard de Vinci*. 3 vols. Paris: A. Hermann, 1906–13.
Fausto, Vittore. *Aristotelis Mechanica Vittoris Fausti industria in pristinum habitum restituta ac latinitate donata*. Paris, 1517.
Favaro, Antonio. 'Delle meccaniche lette in Padova l'anno 1594 da Galileo Galilei.' *Memorie del Reale Istituto Veneto di scienza, lettere ed arti*, 26.5. Venice: Carlo Ferrari, 1899.
– 'Amici e corrispondenti di Galileo Galilei: XL. Giuseppe Moletti.' *Atti del Reale Istituto Veneto di Scienze, Lettere ed Arti* 77 (1917–18): 45–118. Rpt. in *Amici et Corrispondenti di Galileo*. 3 vols. Ed. Paolo Galluzzi. Florence: Salimbeni, 1983. 3:1585–1656.
– 'Giuseppe Moletti.' In *Gli Scienziati Italiani*. ed. A. Mieli. Rome, 1921. 1.1:36–9.
– 'I lettori di matematica nella Università di Padova.' *Memorie e Documenti per la Storia dell' Università di Padova* 1 (1922): 3–70.
– *Galileo Galilei e lo Studio di Padova*. Padua: Antenore, 1966.
Fenlon, Iain. *Music and Patronage in Sixteenth-Century Mantua*. Cambridge: Cambridge Univ. Press, 1980.
Florio, John. *Queen Anna's New World of Words; or, Dictionarie of the Italian and English Tongues*. London, 1611. Rpt. Menston, England: Scolar Press, 1968.
Fredette, Raymond. 'Galileo's "De motu antiquiora."' *Physis* 14 (1972): 321–50.
Gabriel, Astrik L. *A Summary Catalogue of Microfilms of One Thousand Scientific Manuscripts in the Ambrosiana Library, Milan*. Notre Dame, Ind.: Mediaeval Institute, 1968.
Galilei, Galileo. *Opere di Galileo Galilei*. Ed. Antonio Favaro. 23 vols. Florence: Barbèra, 1891–1909.
– 'Delle meccaniche lette in Padova l'anno 1594 da Galileo Galilei.' Ed. Antonio Favaro. *Memorie del Reale Istituto Veneto di Scienze, Lettere ed Arti*, 26.5. Venice: Carlo Ferrari, 1899.

– *Dialogue Concerning the Two Chief World Systems, Ptolomaic and Copernican.* Tr. Stillman Drake. 2nd ed. Berkeley: Univ. California Press, 1967.
Geanakoplos, D.J. 'The Career of the Little-Known Renaissance Greek Scholar Nicholas Leonicus Thomaeus.' *Byzantina* 13 (1985): 355–72.
Giacobbe, Giulio Cesare. 'Il *Commentarium de certitudine mathematicarum disciplinarum* di Alessandro Piccolomini.' *Physis* 14 (1972): 162–93.
– 'Francesco Barozzi e la *Questio di certitudine mathematicarum.*' *Physis* 14 (1972): 357–74.
– 'Alcune cinquecentine riguardanti il processo di rivalutazione epistemologica della matematica nell'ambito della rivoluzione scientifica rinascimentale.' *La Berio* 13 (1973): 7–44.
– 'La riflessione metamatematica di Pietro Catena.' *Physis* 15 (1973): 178–96.
– 'Epigoni nel Seicento della "*Quaestio de certitudine mathematicarum*": Giuseppe Biancani.' *Physis* 18 (1976): 5–40.
– 'Una gesuita progressista nella "Questio de certitudine mathematicarum" rinascimentale: Benito Pereyra.' *Physis* 19 (1977): 51–86.
Grant, Edward. 'Motion in the Void and the Principle of Inertia in the Middle Ages.' *Isis* 55 (1964): 265–92. Rpt. in Grant, *Studies in Medieval Science and Natural Philosophy.*
– 'Aristotle, Philoponus, Avempace, and Galileo's Pisan Dynamics.' *Centaurus* 11 (1965): 79–95. Rpt. in Grant, *Studies in Medieval Science and Natural Philosophy.*
– 'Bradwardine and Galileo: Equality of Velocities in the Void.' *Archive for the History of Exact Sciences* 2 (1965): 344–64. Rpt. in Grant, *Studies in Medieval Science and Natural Philosophy.*
– 'Medieval Explanation and Interpretation of the Dictum that 'Nature Abhors a Vacuum.' *Traditio* 29 (1973): 327–55. Rpt. in Grant, *Studies in Medieval Science and Natural Philosophy.*
– *Much Ado about Nothing: Theories of Space and Vacuum from the Middle Ages to the Scientific Revolution.* Cambridge: Cambridge Univ. Press, 1981.
– *Studies in Medieval Science and Natural Philosophy.* London: Variorum, 1981.
Grendler, Marcella. 'A Greek Collection in Padua: The Library of Gian Vincenzo Pinelli (1535–1601).' *Renaissance Quarterly* 33 (1980): 386–416.
Hall, Bert S. 'Production et diffusion de certains traits de techniques au moyen âge.' In *Les arts mécaniques au moyen âge,* ed. G.H. Allard and S. Lusignan. Montreal: Bellarmine; Paris: Vrin, 1982. Pp. 147–70.
Heath, Sir Thomas L. *A History of Greek Mathematics.* 2 vols. Oxford: Clarendon Press, 1921.
– *Mathematics in Aristotle.* Oxford: Clarendon Press, 1949.

– *The Thirteen Books of Euclid's Elements.* 3 vols. 2nd ed. Cambridge: Cambridge Univ. Press. Rpt. New York: Dover, 1956.

Jope, James. 'Subordinate Demonstrative Science in the Sixth Book of Aristotle's *Physics.*' *Classical Quarterly* 22 (1972): 279–92.

Kennedy, E.S. 'Late Medieval Planetary Theory.' *Isis* 57 (1966): 365–78.

Kristeller, Paul Oskar, F. Edward Cranz, and Virgina Brown, eds. *Catalogus translationum et commentariorum: Medieval and Renaissance Latin Translations and Commentaries.* 7 vols. Washington: Catholic Univ. of America Press, 1960–92.

Laird, W.R. 'The *Scientiae Mediae* in Medieval Commentaries on Aristotle's *Posterior Analytics.*' PhD diss. Univ. Toronto, 1983.

– 'The Scope of Renaissance Mechanics.' *Osiris,* 2nd series 2 (1986): 43–68.

– 'Giuseppe Moletti's "Dialogue on Mechanics" (1576).' *Renaissance Quarterly* 40 (1987): 209–23.

– 'Archimedes among the Humanists.' *Isis* 82 (1991): 629–38.

– 'Patronage of Mechanics and Theories of Impact in Sixteenth-Century Italy.' In *Patronage and Institutions: Science, Technology and Medicine at the European Court, 1550–1750,* ed. Bruce T. Moran. Woodbridge: Boydell Press, 1991. Pp. 51–66.

Leonico Tomeo, Niccolò. *Aristotelis Quaestiones mechanicae.* In *Opuscula nuper in lucem aedita.* Venice, 1525.

Livesey, Steven J. '*Metabasis:* The Interrelationship of the Sciences in Antiquity and the Middle Ages.' PhD diss. Univ. California, Los Angeles, 1982.

Livy. *Ab urbe condita.* Ed. C.F. Walters and R.S. Conway. 5 vols. Oxford: Clarendon Press, 1914.

Lohr, Charles H. 'Renaissance Latin Aristotle Commentaries: Authors C.' *Renaissance Quarterly* 28 (1975): 689–741.

– 'Renaissance Latin Aristotle Commentaries: Authors L–M.' *Renaissance Quarterly* 31 (1978): 532–603.

– 'Renaissance Latin Aristotle Commentaries: Authors Pi–Sm,' *Renaissance Quarterly* 33 (1980): 623–734.

– 'Renaissance Latin Aristotle Commentaries: Authors So–Z.' *Renaissance Quarterly* 35 (1982): 164–256.

– *Latin Aristotle Commentaries II. Renaissance Authors.* Florence: Olschki, 1988.

Long, Pamela O. 'The Contribution of Architectural Writers to a "Scientific" Outlook in the Fifteenth and Sixteenth Centuries.' *Journal of Medieval and Renaissance Studies* 15 (1985): 265–98.

Mahoney, Michael S. 'Mathematics.' In *Science in the Middle Ages,* ed. David C. Lindberg. Chicago: Univ. Chicago Press, 1978. Pp. 145–78.

Maier, Anneliese. 'Diskussionen über das actuell Unendliche in der ersten Hälfte des 14. Jahrhunderts.' *Divus Thomas,* 3rd series 24 (1947): 147–66, 317–37; rpt.

in *Ausgehendes Mittelalter: Gesammelte Aufsätze zur Geistesgeschichte des 14. Jahrhunderts*. Rome: Storia e Letteratura, 1964. 1:41–85, 460–2.

– *Die Vorläufer Galileis*. Rome: Storia e Letteratura, 1949.

– *Zwei Grundprobleme der scholastischen Naturphilosophie*. 2nd ed. Rome: Storia e Letteratura, 1951.

– 'Die naturphilosophie Bedeutung der scholastischen Impetustheorie.' *Scholastik* 30 (1955): 321–43. Rpt. in Maier, *Ausgehendes Mittelalter* 1. Rome: Storia e Letteratura, 1964. Pp. 353–79. Tr. Steven D. Sargent, in *On the Threshold of Exact Science: Selected Writings of Anneliese Maier on Late Medieval Natural Philosophy*. Chicago: Univ. Chicago Press, 1982. Pp. 76–102.

Masotti, Arnaldo. 'Maurolico, Francesco.' In *The Dictionary of Scientific Biography*. 16 vols. New York: Scribner, 1970–80. 9:190–4.

– 'Tartaglia, Niccolò.' In *The Dictionary of Scientific Biography*. 16 vols. New York: Scribner, 1970–80. 13:258–62.

Maurolico, Francesco. *Problemata mechanica cum appendice*. Ed. Silvestro Maurolico. Messina, 1613.

McKirahan, Richard D., Jr. 'Aristotle's Subordinate Sciences.' *British Journal for the History of Science* 11 (1978): 197–220.

Micheli, Gianni. *Le origini del concetto di macchina*. Biblioteca di Physis 4. Leo S. Olschki: Florence, 1995.

Moletti, Giuseppe. *Discorso universale ... nel quale son raccolti, and dichiarati tutti i termini, & tutte le regole appartenenti alla Geografia*. Printed with *La Geografia di Claudio Tolomeo*, tr. Girolamo Ruscelli. Venice, 1561. Ed. Gio. Malombra, Venice, 1564. Printed separately, Venice, 1573.

– *L'Efemeridi di M. Gioseppe Moleto Matematico. Per anni XVIII* [i.e., 1563–80]. Venice, 1563. Extended to 1584, Venice, 1564.

– *De corrigendo Ecclesiastico Calendario libri duo*. Venice, 1580.

– *Tabulae Gregorianae motuum octavae sphaerae*. Venice, 1580.

Malaguzzi, Orazio. *Discorso ... della grandezza de' Stati, Domini de cinque frà li più potenti Regi e Signori dell'Universo*. 1590. = Giuseppe Moletti, 'Discorso che il Re Catolico sia il maggior principe del mondo,' Milan, Biblioteca Ambrosiana MS. P 145 sup., ff. 32r–41v.

Monantheuil, Henri de. *Mechanica Graeca, emendata, Latina facta, et Commentariis illustrata*. Paris, 1599.

Moody, Ernest A. 'Galileo and Avempace: The Dynamics of the Leaning Tower Experiment.' *Journal of the History of Ideas* 12 (1951): 163–93, 375–422. Rpt. in Moody, *Studies in Medieval Philosophy, Science, and Logic*. Berkeley: Univ. California Press, 1975. Pp. 203–85.

– 'Galileo and His Precursors.' In *Studies in Medieval Philosophy, Science, and Logic*. Berkeley: Univ. California Press, 1976. Pp. 393–408.

Moody, Ernest A., and Marshall Clagett, eds. *The Medieval Science of Weights.* Madison: Univ. Wisconsin Press, 1952.

Moraux, Paul. *Les listes anciennes des ouvrages d'Aristote.* Louvain: Éditions Universitaires, 1951.

Murdoch, John E. 'The Medieval Language of Proportions.' In *Scientific Change,* ed. A. C. Crombie. New York: Basic Books, 1963. Pp. 237–71.

– 'Philosophy and the Enterprise of Science in the Later Middle Ages.' In *The Interaction between Science and Philosophy,* ed. Yehuda Elkana. Atlantic Highlands, NJ: Humanities Press, 1974. Pp. 51–74.

– 'From Social to Intellectual Factors: An Aspect of the Unitary Character of Late Medieval Learning.' In *The Cultural Context of Medieval Learning,* ed. John E. Murdoch and Edith D. Sylla. Dordrecht: Reidel, 1975. Pp. 271–339.

– 'Henry of Harclay and the Infinite.' In *Studi sul XIV secolo in memoria Anneliese Maier,* ed. Alfonso Maierù and Agostino Paravicini Bagliani. Rome: Storia e Letteratura, 1981. Pp. 219–61.

Murdoch, John E., and Edith D. Sylla. 'The Science of Motion.' In *Science in the Middle Ages,* ed. David C. Lindberg. Chicago: Univ. Chicago Press, 1978. Pp. 206–64.

Narducci, E. 'Vite Inedite di Matematici Italiani scritte da Bernardino Baldi.' *Bullettino di Bibliografia e di Storia delle Scienze Matematiche e Fisiche* 19 (1886): 625–33.

Ovitt, George, Jr. 'The Status of the Mechanical Arts in Medieval Classifications of Learning.' *Viator* 14 (1983): 89–105.

Pepper, Simon, and Nicholas Adams. *Firearms and Forifications: Military Architecture and Siege Warfare in Sixteenth-Century Italy.* Chicago: Univ. Chicago Press, 1986.

Piccolomini, Alessandro. *De certitudine mathematicarum.* Printed with *In Mechanicas quaestiones Aristotelis paraphrasis.* 2nd ed., Venice, 1565.

– *In Mechanicas quaestiones Aristotelis paraphrasis.* 2nd ed., Venice, 1565.

Pliny. *Historia naturalis.* Ed. and trans. H. Rackham. 10 vols. Loeb Classical Library. London: Heinemann; Cambridge, Mass.: Harvard Univ. Press, 1942.

Plutarch. *Plutarch's Lives.* Ed. and trans. Bernadotte Perrin. 11 vols. Loeb Classical Library. London: Heinemann; New York: Putnam's Sons, 1917.

Proclus. *Procli Diadochi Lycii in primum Euclidis Elementorum commentariorum libri iv.* Tr. Francesco Barozzi. Padua, 1560.

– *Procli Diadochi in primum Euclidis Elementorum librum commentarii.* Ed. Godfried Friedlein. Leipzig, 1873.

– *A Commentary on the First Book of Euclid's Elements.* Tr. Glenn R. Morrow. Princeton: Princeton Univ. Press, 1970; 2nd ed. 1992.

Pseudo-Aristotle. *Mechanical Problems.* Ed. and tr. W.S. Hett, in *Aristotle, Minor*

Works. Loeb Classical Library. London: Heinemann; Cambridge, Mass.: Harvard Univ. Press, 1936.
– *Aristotle, Mechanika*. Ed. Maria Elisabetta Bottecchia. Studia Aristotelica 10. Padua: Antenore, 1982.
Reidy, Martin F. 'Aristotle's Doctrines Concerning Applied Mathematics.' PhD. diss., Univ. Toronto, 1968.
Revelli, Paolo. 'Un trattato geografico-politico de Giuseppe Moleti: "Discorso che il Re Cattolico sia il maggior principe del mondo" [1580–81].' *Aevum* 3 (1927): 417–54.
– *I codici Ambrosiani di contenuto geografica*. Milan, 1929.
Riccoboni, Antonio, *Orationum volumen secundum*. Padua, 1591.
Rivolta, Adolfo. *Catalogo dei codici Pinelliani (latini) dell' Ambrosiana*. Milan: Tipografia Pontificia Archivescovile S. Giuseppe, 1933.
Rose, Paul Lawrence. *The Italian Renaissance of Mathematics*. Geneva: Droz, 1975.
– 'Professors of Mathematics at Padua University 1521–1588.' *Physis* 17 (1975): 300–4.
– 'A Venetian Patron and Mathematician of the Sixteenth Century: Francesco Barozzi (1537–1604).' *Studi Veneziani*, new series 1 (1977): 119–78.
– 'Monte, Guidobaldo, Marchese del.' *The Dictionary of Scientific Biography*. 16 vols. New York: Scribner, 1970–80. 9:487–9.
Rose, Paul Lawrence, and Stillman Drake. 'The Pseudo-Aristotelian *Questions of Mechanics* in Renaissance Culture.' *Studies in the Renaissance* 18 (1971): 65–104.
Sacrobosco, Johannes de. *Tractatus de sphaera*. Ed. and tr. Lynn Thorndike, *The Sphere of Sacrobosco and Its Commentators*. Chicago: Univ. Chicago Press, 1949.
Sargent, Steven D., tr. *On the Threshold of Exact Science: Selected Writings of Anneliese Maier on Late Medieval Natural Philosophy*. Chicago: Univ. Chicago Press, 1982.
Schmitt, Charles B. 'Experimental Evidence for and against a Void: The Sixteenth-Century Arguments.' *Isis* 58 (1967): 352–66.
Schramm, M. 'The Mechanical Problems of the *Corpus Aristotelicum*, the *Elementa Iordani super Demonstrationem Ponderum*, and the Mechanics of the Sixteenth Century.' In *Atti del Primo Convegno Internazionale di Ricognizione delle Fonti per la Storia della Scienza Italiana: I Secoli XIV–XVI*, ed. Carlo Maccagni. Florence: Barbèra, 1967. Pp. 151–63.
Settle, Thomas B. 'Galileo and Early Experimentation.' In *Springs of Scientific Creativity: Essays on Founders of Modern Science*, ed. Rutherford Aris, H. Ted Davis, and Roger H. Stuewer. Minneapolis: Univ. Minnesota Press, 1983. Pp. 3–20.
Suter, Rufus. 'The Scientific Work of Alessandro Piccolomini.' *Isis* 60 (1969): 210–22.
Tagmann, Pierre M., and Michael Fink. 'Rovigo, Francesco (Franceschino).' *The*

New Grove Dictionary of Music and Musicians. London: Macmillan, 1980. 16:279–80.

Tartaglia, Niccolò. *Nova Scientia.* Venice, 1537.

– *Quesiti et inventioni diverse.* Venice, 1554; facs. ed. Brescia: Ateneo di Brescia, 1959.

Thomas Bradwardine. *Tractatus de proportionibus velocitatum in motibus.* Ed. and tr. H. Lamar Crosby, Jr. Madison: Univ. Wisconsin Press, 1961.

Venturi, Giambatista. *Memorie e lettere inedita finora o disperse di Galileo Galilei.* Modena, 1818.

Vitruvius. *I dieci libri dell'Architettura di Vitruvio.* Trans. Danielo Barbaro. Venice, 1556.

– *De architectura libri X.* Ed. and trans. Frank Granger. 2 vols. Loeb Classical Library. London: Heinemann; Cambridge, Mass.: Harvard Univ. Press, 1956.

– *Vitruve, De l'Architecture livre X.* Ed. and tr. Louis Callebat and Philippe Fleury. Paris: Belles Lettres, 1986.

Wallace, William A. *Galileo and His Sources: The Heritage of the Collego Romano in Galileo's Science.* Princeton: Princeton Univ. Press, 1984.

Weisheipl, James A. 'Natural and Compulsory Movement.' *New Scholasticism* 29 (1955): 50–81. Rpt. in James A. Weisheipl, OP, *Nature and Motion in the Middle Ages,* ed. William E. Carroll. Washington: Catholic Univ. America Press, 1985. Pp. 25–48.

– 'Classifications of the Sciences in Medieval Thought.' *Mediaeval Studies* 27 (1965): 54–90.

– 'Motion in a Void: Aquinas and Averroes.' In *St. Thomas Aquinas 1274–1974: Commemorative Studies.* 2 vols., ed. A.A. Maurer. Toronto: Pontifical Institute of Mediaeval Studies, 1974. 1:467–88. Rpt. in Weisheipl, *Nature and Motion in the Middle Ages,* ed. William E. Carroll. Washington: Catholic Univ. America Press, 1985. Pp. 121–42.

Westman, Robert S. 'The Astronomer's Role in the Sixteenth Century: A Preliminary Study.' *History of Science* 18 (1980): 105–45.

Whitney, Elspeth. 'Paradise Restored: The Mechanical Arts from Antiquity through the Thirteenth Century.' *Transactions of the American Philosophical Society* 80, part 1. 1990.

Wilkinson, Catherine. 'Renaissance Treatises on Military Architecture and the Science of Mechanics.' In *Les traités d'architecture de la renaissance,* ed. Jean Guillaume. Paris: Picard, 1988. Pp. 467–76.

Index